AVIATION PSYCHOLOGY and HUMAN FACTORS

AVIATION PSYCHOLOGY and HUMAN FACTORS

Monica Martinussen
David R. Hunter

CRC Press
Taylor & Francis Group
Boca Raton   London   New York

CRC Press is an imprint of the
Taylor & Francis Group, an **informa** business

CRC Press
Taylor & Francis Group
6000 Broken Sound Parkway NW, Suite 300
Boca Raton, FL 33487-2742

© 2010 by Taylor and Francis Group, LLC
CRC Press is an imprint of Taylor & Francis Group, an Informa business

Printed in the United States of America on acid-free paper
10 9 8 7 6 5 4 3 2 1

International Standard Book Number: 978-1-4398-0843-6 (Hardback)

First published by Fagbokforlaget, Norway, in 2008.

Library of Congress Cataloging-in-Publication Data

Martinussen, Monica.
   Aviation psychology and human factors / Monica Martinussen, David R. Hunter.
     p. cm.
   Includes bibliographical references and index.
   ISBN 978-1-4398-0843-6 (hardcover : alk. paper)
   1. Aviation psychology. 2. Flight--Physiological aspects. I. Hunter, David R. (David Robert), 1945- II. Title.

TL555.M365 2010
155.9'65--dc22
                                    2009026281

Visit the Taylor & Francis Web site at
http://www.taylorandfrancis.com

and the CRC Press Web site at
http://www.crcpress.com

Contents

Preface

In a lecture at the University of Lille on December 7, 1854, Louis Pasteur noted that in the field of observational science, "le hasard ne favorise que les esprits préparés" ("fortune favors the prepared mind"). We would suggest that this statement is equally true in the field of aviation. However, reversing the statement might make it even more cogent because, for aviators, *misfortune punishes the unprepared mind.* To survive in the demanding domain of aviation, one needs to approach the task fully prepared. This means having a comprehensive knowledge not only of weather, aerodynamics, propulsion, navigation, and all the other technical disciplines, but also of what is simultaneously the most fragile and most resilient, the most unreliable and the most adaptable component: the human being. The study of aviation psychology can provide some of that knowledge and offer better preparation for the demands that a life or an hour in aviation will make.

This book is about applied psychology. Specifically, it is concerned with the application of psychological principles and techniques to the specific situations and problems of aviation. The book is meant to inform the student of psychology about how the discipline is applied to aviation; even more, it is meant to inform the student of aviation about how psychology can be used to address his or her concerns. We attempt to maintain this balance of perspectives and needs throughout the book; however, when we slip, we do so in favor of the student of aviation. Many books have been written by psychologists for psychologists. Few, if any, books have been written by psychologists for pilots. It is to this neglected segment that we offer the main thrust of this work.

The overall goal of the book is to make pilots aware of the benefits of psychology and its application for improving aviation operations, as well as to provide specific information that pilots can use in their daily operations. In addition to making pilots aware of the benefits of psychology, the book should also make pilots informed consumers of psychological research and studies so that they may better evaluate and implement future products in the field of aviation psychology.

We would like to express our gratitude to colleagues and students who have read parts of the book or the complete manuscript and provided us with valuable feedback. In particular, we wish to mention Dr. Kjell Mjøs and military psychologist Live Almås-Sørensen. The book was first published in Norwegian with the title *Luftfartspsykologi* (2008) by the publisher Fagbokforlaget. Special thanks are offered to Martin Rydningen for helping us with the translation process, both from English to Norwegian and from Norwegian to English.

Monica Martinussen
David Hunter

About the Authors

Monica Martinussen is currently professor of psychology in the medical school of the University of Tromsø, Norway, and is a licensed psychologist. Dr. Martinussen conducted her doctoral research in the area of pilot selection and has been engaged in research on that topic for many years, both for the Norwegian Air Force and at the University of Tromsø. Her research interests include research methods and psychometrics, aviation psychology, and work and organizational psychology. Dr. Martinussen is a member of the board of directors of the European Association of Aviation Psychology.

David Hunter has more than 30 years experience as an aviation psychologist. He is a former military helicopter pilot with combat experience in Southeast Asia, and he holds a commercial pilot license for fixed-wing aircraft and for helicopters. Dr. Hunter has conducted research on the use of computer-based tests for the selection of pilots for the U.S. Air Force and the UK Royal Air Force, and he has served as an advisor for the human factors design of new aviation systems for the U.S. Army. He also served as the principal scientist for human performance with the Federal Aviation Administration, where he conducted research and managed programs to improve safety among general aviation pilots.

1 Introduction

1.1 WHAT IS AVIATION PSYCHOLOGY?

Because the primary target of this book is the student of aviation, rather than the student of psychology, it seems prudent to begin with a few definitions. This will set some bounds for our discussions and for the reader's expectations. The title of the book includes two key terms: "aviation psychology" and "human factors." We included both these terms because they are often used interchangeably, although that is a disservice to both disciplines. Although we will touch on some of the traditional areas of human factors in the chapter on the design of aviation systems, our primary focus is on aviation psychology. Therefore, we will dwell at some length on what we mean by that particular term.

Psychology* is commonly defined as the science of behavior and mental processes of humans, although the behavior of animals is also frequently studied—usually as a means to understand human behavior better. Within this broad area, there are numerous specialties. The American Psychological Association (APA), the largest professional organization of psychologists, lists over 50 divisions, each representing a separate aspect of psychology. These include several divisions concerned with various aspects of clinical psychology along with divisions concerned with such diverse issues as consumer behavior, school psychology, rehabilitation, the military, and addiction. All of these are concerned with understanding how human behavior and mental processes influence or are influenced by the issues of their particular domain.

Clearly, psychology covers a very broad area: Literally, any behavior or thought is potential grist for the psychologist's mill. To understand exactly what this book will cover, let us consider what we mean by aviation psychology. Undoubtedly, students of aviation will know what the first part of the term means, but what is included under psychology, and why do we feel justified, even compelled, to distinguish between aviation psychology and the rest of the psychological world?

First, let us immediately dismiss the popular image of psychology. We do not include in our considerations of aviation psychology reclining on a couch recounting our childhood and the vicissitudes of our emotional development. That popular image of psychology belongs more to the area of clinical psychology, or perhaps even psychoanalysis. Although clinical psychology is a major component of the larger field of psychology, it has little relevance to aviation psychology. That is not to say that pilots and others involved in aviation are not subject to the same mental foibles and afflictions that beset the rest of humanity. Neither would we suggest that aspects

---

* The term "psychology" is derived from the Greek word *psyche*, meaning both butterfly and soul. Psychology was first used as part of a course title for lectures given in the sixteenth century by Philip Melanchton.

of the human psyche usually addressed in a clinical setting could have no influence on human performance in an aviation setting. Quite the opposite, we assert that all aspects of the mental functioning of pilots, maintenance personnel, air traffic controllers, and the supporting cadre inescapably influence behavior for better or worse.

Rather, we wish to dissociate aviation psychology from the psychotherapeutic focus of traditional clinical psychology. Aviation psychology may concern itself with the degree of maladaptive behavior evidenced by excessive drinking or with the confused ideation associated with personality disorders. However, it does so for the purpose of understanding and predicting the effects of those disorders and behaviors on aviation-related activities, rather than for the purpose of effecting a cure.

Ours is a much more basic approach. We are concerned not only with the behavior (what people do) and ideation (what people think) of those with various mental disturbances, but also with how people in general behave. Psychology at its most inclusive level is the study of the behavior of all people. Psychology asks why, under certain conditions, people behave in a certain way, and under different conditions they behave in a different way. How do prior events, internal cognitive structures, skills, knowledge, abilities, preferences, attitudes, perceptions, and a host of other psychological constructs (see the later discussion of constructs and models) influence behavior? Psychology asks these questions, and psychological science provides the mechanism for finding answers. This allows us to understand and to predict human behavior.

We may define aviation psychology as the study of individuals engaged in aviation-related activities. The goal of aviation psychology, then, is to understand and to predict the behavior of individuals in an aviation environment. Being able, even imperfectly, to predict behavior has substantial benefits. Predicting accurately how a pilot will react (behave) to an instrument reading will allow us to reduce pilot error by designing instruments that are more readily interpretable and that do not lead to incorrect reactions. Predicting how a maintenance technician will behave when given a new set of instructions can lead to increased productivity through reduction of the time required to perform a maintenance action. Predicting how the length of rest breaks will affect an air traffic controller faced with a traffic conflict can lead to improved safety. Finally, predicting the result of a corporate restructuring on the safety culture of an organization can identify areas in which conflict is likely to occur and areas in which safety is likely to suffer.

From this general goal of understanding and predicting the behavior of individuals in the aviation environment, we can identify three more specific goals: first, to reduce error by humans in aviation settings; second, to increase the productivity; and third, to increase the comfort of both the workers and their passengers. To achieve these goals requires the coordinated activities of many groups of people. These include pilots, maintenance personnel, air traffic control operators, the managers of aviation organizations, baggage handlers, fuel truck drivers, caterers, meteorologists, dispatchers, and cabin attendants. All of these groups, plus many more, have a role in achieving the three goals of safety, efficiency, and comfort. However, because covering all these groups is clearly beyond the scope of a single book, we have chosen to focus on the pilot, with only a few diversions into the activities of the other groups. Another reason for choosing pilots is that the majority of research has been conducted on pilots. This is slowly changing, and more research is being conducted using air traffic controllers, crew members, and other occupational groups involved in aviation.

In this, we enlist contributions from several subdisciplines within the overall field of psychology. These include engineering psychology and its closely related discipline of human factors, personnel psychology, cognitive psychology, and organizational psychology. This listing also matches, to a fair degree, the order in which we develop our picture of aviation psychology—moving from fairly basic considerations of how the operator interacts with his or her aircraft (the domain of engineering psychology and human factors) through considerations of how best to select individuals to be trained as pilots (the domain of personnel and training psychology). Cognitive psychology also contributes to our understanding of how individuals learn new tasks, along with providing us information on how best to structure jobs and training so that they match the cognitive structure of the learner. Finally, from organizational psychology we learn how the structure and climate of an organization can contribute to issues such as safety through the expectations for behavior fostered among members of the organizations, as well as by the reporting and management structure that the corporate executives put in place.

Although aviation psychology draws heavily upon the other disciplines of psychology, those other disciplines are also heavily indebted to aviation psychology for many of their advances, particularly in the area of applied psychology. This is due primarily to the historic ties of aviation psychology to military aviation. For a number of reasons to be discussed in detail later, aviation—and pilots in particular—have always been a matter of very high concern to the military. Training of military pilots is an expensive and lengthy process, so considerable attention has been given since World War I to improving the selection of these individuals so as to reduce failures in training—the provenance of personnel and training psychology.

Similarly, the great cost of aircraft and their loss due to accidents contributed to the development of engineering psychology and human factors. Human interaction with automated systems, now a great concern in the computer age, has been an issue of study for decades in aviation, beginning from the introduction of flight director systems and in recent years the advanced glass cockpits. Much of the research developed in an aviation setting for these advanced systems is equally germane to the advanced displays and controls that will soon appear in automobiles and trucks.

In addition, studies of the interaction of crew members on airliner flight decks and the problems that ensue when one of the other crew members does not clearly assert his or her understanding of a potentially hazardous situation has led to the development of a class of training interventions termed crew resource management (CRM). After a series of catastrophic accidents, the concept and techniques of CRM were developed by the National Aeronautics and Space Administration (NASA) and the airline industry to ensure that a crew operates effectively as a team. Building upon this research base from aviation psychology, CRM has been adapted for other settings, such as air traffic control centers, medical operating rooms, and military command and control teams. This is a topic we will cover in much more detail in a later chapter.

1.2   WHAT IS RESEARCH?

Before we delve into the specifics of aviation psychology, however, it may be worthwhile to consider in somewhat greater detail the general field of psychology. As noted earlier, psychology is the science of behavior and mental process. We describe it as a science because psychologists use the scientific method to develop their knowledge of behavior and mental process. By agreeing to accept the scientific method as the mechanism by which truth will be discovered, psychologists bind themselves to the requirements to test their theories using empirical methods, and they modify or reject those that are not supported by the results. The APA defines scientific method as "the set of procedures used for gathering and interpreting objective information in a way that minimizes error and yields dependable generalizations."*

> For a discussion of the "received view" of the philosophy of science, see Popper (1959) and Lakatos (1970). According to this view, science consists of bold theories that outpace the facts. Scientists continually attempt to falsify these theories but can never prove them true. For discussions on the application of this philosophy of science to psychology, see Klayman and Ha (1987), Poletiek (1996), and Dar (1987).

At a somewhat less lofty level, scientific method consists of a series of fairly standardized steps, using generally accepted research procedures:

- Identify a problem and formulate a hypothesis (sometimes called a theory).
- Design an experiment that will test the hypothesis.
- Perform the experiment, typically using experimental and control groups.
- Evaluate the results from the experiment to see if the hypothesis was supported.
- Communicate the results.

For example, a psychologist might observe that a large number of pilot trainees fail during their training (the *problem*). The psychologist might form a hypothesis, possibly incorporating other observations or information, that the trainees are failing because they are fatigued and that the source of this fatigue is a lack of sleep. The psychologist might then formally state her hypothesis that the probability of succeeding in training is directly proportional to the number of hours of sleep received (the *hypothesis*). The psychologist could then design an experiment to test that hypothesis.

In an ideal experiment (not likely to be approved by the organization training the pilots), a class of incoming trainees would be randomly divided into two groups. One group would be given X hours of sleep, and the other group would be given Y hours of sleep, where Y is smaller than X (*design* the experiment). The trainees would

---

be followed through the course and the numbers of failures in each group recorded (*perform* the experiment). The results could then be analyzed using statistical methods (*evaluate* the results) to determine whether, as predicted by the psychologist's hypothesis, the proportion of failures in group X was smaller than the proportion of failures in group Y.

If the difference in failure rates between the two groups was in the expected direction and if it met the generally accepted standards for statistical significance, then the psychologist would conclude that her hypothesis was supported, and she would indicate this in her report (*communicate* the results). If the data she collected from the experiment did not support the hypothesis, then she would have to reject her theory or modify it to take the results of the experiment into account.

Like research in other fields, psychological research must meet certain criteria in order to be considered scientific. The research must be

falsifiable; the hypothesis or theory must be stated in a way that makes it possible to reject it. If the hypothesis cannot be tested, then it does not meet the standards for science.

replicable; others should be able to repeat a study and get the same results. It is for this reason that reports of studies should provide enough detail for other researchers to repeat the experiment.

precise; hypotheses must be stated as precisely as possible. For example, if we hypothesize that more sleep improves the likelihood of completing pilot training, but only up to some limit (that is, trainees need 8 hours of sleep, but additional hours beyond that number do not help), then our hypothesis should explicitly state that relationship. To improve precision, operational definitions of the variables should be included that state exactly how a variable is measured. Improved precision facilitates replication by other researchers.

parsimonious; researchers should apply the simplest explanation possible to any set of observations. This principle, sometimes called Occam's razor, means that if two explanations equally account for an observation, then the simpler of the two should be selected.

1.3   GOALS OF PSYCHOLOGY

Describe. Specify the characteristics and parameters of psychological phenomena more accurately and completely. For example, studies have been conducted of human short-term memory that very accurately describe the retention of information as a function of the amount of information to be retained.

Predict. Predicting what people will do in the future, based on knowledge of their past and current psychological characteristics, is a vital part of many aviation psychology activities. For example, accurately predicting who will complete pilot training based on knowledge of their psychological test scores is important to the organization performing the training. Likewise, predicting who is more likely to be in an aircraft accident based on psychological test scores could also be valuable information for the person involved.

Understand. This means being able to specify the relationships among variables—in plain language, knowing the "how" and "why" drives psychologists and nonpsychologists alike. Once we understand, we are in a position to predict and to influence.

Influence. Once we have learned why a person fails in training or has an accident, we may be able to take steps to change the outcome. From our earlier example, if we know that increasing the amount of sleep that trainees receive improves their likelihood of succeeding in training, then we almost certainly will wish to change the training schedule to ensure that everyone gets the required amount of sleep every night.

Psychology can also be broken down into several different general approaches. These approaches reflect the subject matter under consideration and, to a large degree, the methods and materials used. These approaches include:

A *behaviorist* approach looks at how the environment affects behavior.

A *cognitive* approach studies mental processes and is concerned with understanding how people think, remember, and reason.

A *biological* approach is concerned with the internal physiological processes and how they influence behavior.

A *social* approach examines how we interact with other people and emphasizes the individual factors that are involved in social behavior, along with social beliefs and attitudes.

A *developmental* approach is primarily interested in emotional development, social development, and cognitive development, including the interactions among these three components.

A *humanistic* approach focuses on individual experiences, rather than on people in general.

The delineation of these six approaches may suggest more homogeneity than actually exists. Although some psychologists remain exclusively within one of these approaches (physiological psychologists are perhaps the best example), for the most part psychologists take a more eclectic view—borrowing concepts, methods, and theories from among the six approaches as it suits their purpose. Certainly, it would be very difficult to classify aviation psychologists into one of these six approaches.

1.4   MODELS AND PSYCHOLOGICAL CONSTRUCTS

The rules of science are met in other disciplines (e.g., chemistry, physics, or mathematics) through the precise delineation of predecessors, actions, conditions, and outcomes. In chemistry, for example, this is embodied in the familiar chemical equation depicting the reaction between two or more elements or compounds. The chemical equation for the generation of water from hydrogen and oxygen is unambiguous: $2H + O = H_2O$. That is, two hydrogen atoms will combine with one oxygen atom to form one molecule of water.

This is a simple, but powerful model that lets chemists understand and predict what will happen when these two elements are united. It also provides a very precise definition of the model, which allows other scientists to test its validity. For example, a scientist might ask: Are there any instances in which H_3O is produced? Clearly, the production of a model such as this is a very desirable state and represents the achievement of the goals, in the chemical domain, that were listed for psychology earlier. Although psychology cannot claim to have achieved the same levels of specificity as the physical sciences, great progress has nevertheless been made in specifying the relationships among psychological variables, often at a quantitative level. However, the level of specificity at present generally is inversely related to the complexity of the psychological phenomenon under investigation.

Some of the earliest work in psychology dealt with psychophysics—generally including issues such as measurement of "just noticeable differences" (JNDs) in the tones of auditory signals or the weights of objects. In Leipzig, Germany, Ernst Weber (1795–1878) discovered a method for measuring internal mental events and quantifying the JND. His observations are formulated into an equation known as *Weber's law,* which states that the just noticeable difference is a constant fraction of the stimulus intensity already present (Corsini, Craighead, and Nemeroff 2001).

More recent efforts have led to the development of several equations describing psychological phenomena in very precise models. These include Fitts's law (Fitts 1954), which specifies that the movement time (e.g., of a hand to a switch) is a logarithmic function of distance when target size is held constant, and that movement time is also a logarithmic function of target size when distance is held constant. Mathematically, Fitts's law is stated as follows:

$$MT = a + b \log_2(2A/W)$$

where

MT = time to complete the movement
a, b = parameters, which vary with the situation
A = distance of movement from start to target center
W = width of the target along the axis of movement

Another such example is Hick's law, which describes the time it takes a person to make a decision as a function of the possible number of choices (Hick 1952). This law states that, given n equally probable choices, the average reaction time (T) to choose among them is

$$T = b\log_2(n + 1)$$

This law can be demonstrated experimentally by having a number of buttons with corresponding light bulbs. When one light bulb is lit randomly, the person must press the corresponding button as quickly as possible. By recording the reaction time, we can demonstrate that the average time to respond varies as the log of the number of light bulbs. Although a seemingly trivial statement of relationships, Hick's and

Fitts's laws are considered in the design of menus and submenus used in a variety of aviation and nonaviation settings (Landauer and Nachbar 1985).

WHAT IS A MODEL?

A model is a simplified representation of reality. It can be a physical, mathematical, or logical representation of a system, entity, phenomenon, or process. When we talk about a psychological model, we are usually referring to a statement, or a series of statements, about how psychological constructs are related or about how psychological constructs influence behavior. These models can be very simple and just state that some things seem to be related. For an example, see the description of the SHEL model.

On the other hand, the model could be quite complex and make specific quantitative statements about the relationships among the constructs. For an example of this type of model, see the weather modeling study in which a mathematical modeling technique is used to specify how pilots combine weather information.

Other models, such as those of human information processing or aeronautical decision making, make statements about how information is processed by humans or how they make decisions. A good model allows us to make predictions about how changes in one part of the model will affect other parts.

Clearly, from some psychological research, very precise models may be constructed of human sensory responses to simple stimuli. Similarly, early work on human memory established with a fairly high degree of specificity the relationship between the position of an item in a list of things to be remembered and the likelihood of its being remembered (Ebbinghaus, 1885, as reprinted in Wozniak 1999).

In addition to highly specific, quantitative models, psychologists have also developed models that specify qualitative or functional relationships among variables. Some models are primarily descriptive and make no specific predictions about relationships among variables other than to suggest that a construct exists and that, in some unspecified way, it influences another construct or behavior. Some models propose a particular organization of constructs or a particular flow of information or events. The predicted relationships and processes of those models may be subject to empirical tests to assess their validity—a very worthwhile characteristic of models. Of particular interest* to the field of aviation psychology are models that deal with

- general human performance;
- skill acquisition and expertise development;

---

* These are but a sampling of the many models currently available. For more information, consult Foyle et al. (2005), Wickens et al. (2003), or Wickens and Holland (2000). An extensive review of human performance models is also available from Leiden et al. (2001), who include task network, cognitive, and vision models. Table 1 in the report by Isaac et al. (2002) also provides a comprehensive listing of models.

- human information processing;
- accident etiology; and
- decision making (specifically, aeronautical decision making).

1.5 HUMAN PERFORMANCE MODELS

One of the more widely used models of human performance is the SHEL model, originated by Edwards (1988) and later modified by Hawkins (1993). The SHEL model consists of the following elements:

- S—software: procedures, manuals, checklists, and literal software;
- H—hardware: the physical system (aircraft, ship, operating suite) and its components;
- E—environment: the situation in which the other elements (L, H, and S) operate, including working conditions, weather, organizational structure, and climate; and
- L—liveware: the people (pilots, flight attendants, mechanics, etc.).

This model is typically depicted as shown in Figure 1.1, which highlights the interrelationships of the S, H, and L components and their functioning within the environment (E).

Although this model is useful at an overall conceptual level, it is pedagogic rather than prescriptive. That is, it serves to help educate people outside the disciplines of psychology and human factors about the interactions and dependencies of the SHEL elements. However, it makes no specific statements about the nature of those interactions and no quantifiable predictions about the results of disruptions. Clearly, this is a simple model of a very complex situation. As such, it has very little explanatory power, although it does serve a general descriptive function.

Despite, or perhaps because of, its simplicity, the SHEL model has proven to be a very popular model within human factors, and it is frequently used to explain

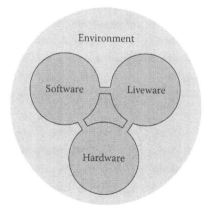

FIGURE 1.1 SHEL model.

concepts relating to the interdependencies of these four elements. It is cited, for example, by the Global Accident Information Network (GAIN 2001) in its chapter on human factors to illustrate the continuous interaction among the elements. It is also referenced in the UK Civil Aviation Authority (CAA 2002) publication on human factors in aircraft maintenance. The SHEL model has also been used extensively outside aviation, particularly in the field of medicine (cf. Bradshaw 2003; Molloy and O'Boyle 2005).

1.5.1 Development of Expertise

How a pilot develops from a novice to an expert is clearly an issue of keen interest to psychologists because of the general agreement that experts are safer than novices (an assumption that should not be accepted without question). Accordingly, several models have been utilized to help understand this process. Some of these models are taken from the general psychological literature on expertise development, and some are specifically adapted to the aviation setting.

One model from the general literature on expertise development specifies five developmental levels—from novice to expert—that reflect an increasing capacity to internalize, abstract, and apply rules (Dreyfus and Dreyfus 1986):

A *novice* learns basic facts, terminology, and rules and how they are applied in well-defined circumstances.

An *advanced beginner* begins to develop a feel for rules through repeated practical application. The student begins to understand the use of concepts and rules in situations that are similar to those in prior examples.

Competence means a deep-enough understanding of the rules to know when they are applicable and how to apply them in novel situations.

Proficiency indicates a refined and internalized sense of the rules.

An *expert* produces increasingly abstract representations and is able to map novel situations to the internalized representations.

This model is primarily descriptive in that it makes no specific predictions regarding the transitions between states, other than the general suggestion that there is an increasing capacity to internalize, abstract, and apply rules. It leaves unanswered questions such as how the process might be accelerated or specifically how to measure competence at each of the hypothesized stages.

According to Fitts (1954; Fitts and Posner 1967), there are three phases to skill acquisition:

- The *cognitive* phase is characterized by slow, declarative learning of primarily verbal information.
- The *associative* phase is characterized by the detection and elimination of errors in performance and the strengthening of connections.
- The *autonomous* phase is characterized by automated and rapid performance, requiring less deliberate attention and fewer resources.

Arguably, this model, like that of Dreyfus and Dreyfus, is primarily descriptive. However, Anderson (1982) has provided a quantitative formulation of the three-phase skill acquisition model. This quantitative model (ACT-R) can be used to make specific quantitative predictions, and as such is subject to more rigorous evaluation than the simply descriptive models. In addition, although originally intended to apply to motor learning, the model is also applicable to cognitive skill acquisition (VanLenh 1996), thus broadening its applicability.

1.6 MODELS OF HUMAN INFORMATION PROCESSING

Perhaps the best known of the human-information-processing models is that proposed by Wickens (Wickens and Holland 2000), which draws heavily upon previous research on human memory (Baddeley 1986), cognition (Norman and Bobrow 1975), and attention (Kahneman 1973). The Wickens model is characterized by the presence of discrete stages for the processing of information, the provision of both a working memory and a long-term memory, and a continuous feedback stream. The provision of an attention resource component is also notable because it implies the notion of a limited attention store, which the human must allocate among all the ongoing tasks. Hence, the attention has a selective nature.

Clearly, this model is more sophisticated in its components and proposed interrelationships than the more descriptive models considered earlier. This level of sophistication and the wealth of detail provide ample opportunity for the evaluation of the validity of this model experimentally. This also makes it a useful tool for understanding and predicting human interaction with complex systems.

1.7 MODELS OF ACCIDENT CAUSATION

The predominant model of accident causation is the Reason (1990) model or, as it is sometimes called, the "Swiss-cheese" model. Because it is so widespread in regard to aviation safety, it will be described at length in the later chapter on safety and hence will not be described in detail here. For present, let us simply note that the Reason model might be properly described as a process model, somewhere midway between the purely descriptive models, like Dreyfus and Dreyfus, and the highly structured model of Wickens. It describes the process by which accidents are allowed or prevented, but also hypothesizes a rather specific hierarchy and timetable of events and conditions that lead to such adverse events.

Moving away from the individual person, there are also models that treat the relationships of organizations and the flow of information and actions within organizations. The current term for this approach is "safety management systems," and it has been adopted by the International Civil Aviation Organization (ICAO 2005) and by all the major Western regulatory agencies, including, for example, the U.S. Federal Aviation Administration, the UK Civil Aviation Authority (CAA 2002), Transport Canada (2001), and the Australian Civil Aviation Safety Authority (CASA 2002).

1.8 MODELS OF AERONAUTICAL DECISION MAKING (ADM)*

Following a study by Jensen and Benel (1977) that found that poor decision making was associated with about half of fatal general aviation accidents, a great deal of interest developed in understanding how pilots make decisions and how that process might be influenced. This interest led to the development of a number of prescriptive models that were based primarily on expert opinion. One such example is the "I'M SAFE" mnemonic device, which serves to help pilots remember to consider the six elements indicated by the IMSAFE letters: illness, medication, stress, alcohol, fatigue, and emotion.

The DECIDE model (Clarke 1986) could be considered both descriptive and prescriptive. That is, it not only describes the steps that a person takes in deciding on a course of action, but also can be used as a pedagogic device to train a process for making decisions. The DECIDE model consists of the following steps:

- D—Detect: the decision maker detects a change that requires attention.
- E—Estimate: the decision maker estimates the significance of the change.
- C—Choose: the decision maker chooses a safe outcome.
- I—Identify: the decision maker identifies actions to control the change.
- D—Do: the decision maker acts on the best options.
- E—Evaluate: the decision maker evaluates the effects of the action.

In an evaluation of the DECIDE model, Jensen (1988) used detailed analyses of accident cases to teach the DECIDE model to 10 pilots. Half of the pilots received the training, while the other half served as a control group. Following the training, the pilots were assessed in a simulated flight in which three unexpected conditions occurred, requiring decisions by the pilots. A review of the experimental flights indicated that all of the experimental group members who chose to fly (four of five) eventually landed safely. All of the control group members who chose to fly (three of five) eventually crashed. Although the very small sample size precluded the usual statistical analysis, the results suggest some utility for teaching the model as a structured approach to good decision making.

As a result of this and other studies conducted at Ohio State University, Jensen and his associates (Jensen 1995, 1997; Kochan et al. 1997) produced the general model of pilot judgment shown in Figure 1.2 and the overarching model of pilot expertise shown in Figure 1.3.

This overall model of pilot expertise was formulated based on four studies of pilot decision making. These studies began with a series of unstructured interviews of pilots who, on the basis of experience and certification, were considered experts in the area of general aviation. The interviews were used to identify and compile characteristics of these expert pilots. Successive studies were used to identify the salient characteristics further, culminating in the presentation to the pilots of a plausible general aviation flight scenario using a verbal protocol methodology. The results from the final study, in combination with the earlier interviews, suggested that, when

---

* O'Hare (1992) provides an in-depth review of the multiple models of ADM.

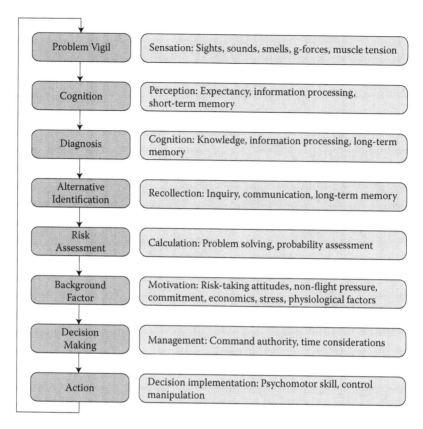

FIGURE 1.2 General model of pilot judgment.

compared to competent pilots, expert pilots tended to (1) seek more quality information in a more timely manner, (2) make more progressive decisions to solve problems, and (3) communicate more readily with all available resources (Kochan et al. 1997).

A somewhat different formulation of the ADM process was offered by Hunter (2002), who suggested that decision making is affected by

- pilot's knowledge;
- pilot's attitudes and personality traits;
- pilot's ability to find and use information effectively;
- quality, quantity, and format of available information;
- pilot's ability to deal with multiple demands;
- pilot's repertoire of possible responses;
- capabilities of aircraft and systems;
- available outside support; and
- outcomes of earlier decisions.

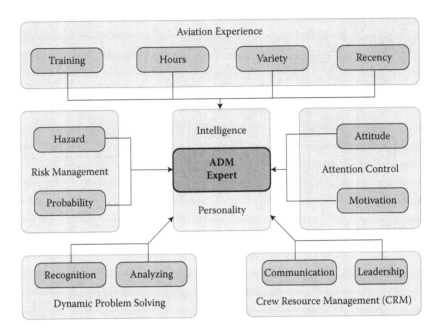

FIGURE 1.3 ADM expertise model.

Hunter combined these elements into the sequence of events and influences into the somewhat more general model of performance depicted in Figure 1.4. This model shares some aspects of Jensen's model, but asserts a greater level of detail. This greater level of detail allows more precise tests of the model to be conducted. For example, Hunter's model suggests that recognition is separate from interpretation and that those stages of processing are influenced by different pilot characteristics. Knowledge (memory) is involved in both stages; however, aspects of the pilot such as personality and risk tolerance have an impact on the interpretation, but not the recognition stages. Experiments could thus be devised to test the predictions of this model.

In a different approach to understanding how pilots make decisions, Hunter, Martinussen, and Wiggins (2003) used a linear modeling technique to examine the weather-related decision-making processes of American, Norwegian, and Australian pilots. In this study, pilots were asked to assign a comfort rating to each of 27 weather scenarios, flown over three different routes. These data were then used to develop individual regression equations* for each pilot that described how each individual pilot combined information about weather conditions (cloud ceiling, visibility, and amount and type of precipitation) to make his or her comfort rating.

Examination of the weights that the pilots used in combining the information allowed Hunter et al. to conclude that pilots among these diverse groups used a

---

* A regression equation is a mathematical equation that shows how information is combined by using weights assigned to each salient characteristic. The general form of the equation is $Y = b_1 x_1 + b_2 x_2$, $+ \ldots + c$, where b1, b2, etc. are the weights applied to each characteristic. Most introductory texts on statistics will include a discussion on linear regression. For a more advanced, but still very readable description of the technique, see Licht (2001).

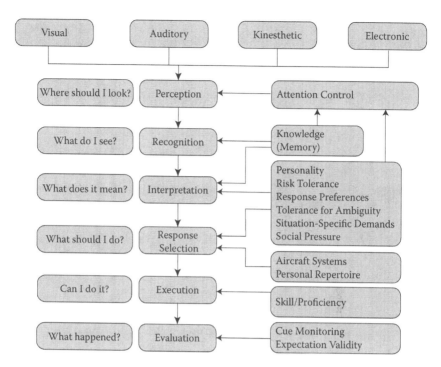

FIGURE 1.4   Hunter's performance model.

consistent weather decision model. For each group, a compensatory model was favored over a noncompensatory model. This means that the pilots were allowing a high score on one of the variables (e.g., ceiling) to compensate for a low score on another of the variables (e.g., visibility). In practical terms, this means that pilots might take off on a flight under potentially hazardous conditions (low visibility, but high ceiling) because of the way the information was combined.

Typically, psychological models become less quantitative and more qualitative as they attempt to account for more complex behavior. For example, contrast the very precise equations that relate response time to number of choices to the models that attempt to account for human information processing. However, the study by Hunter and colleagues demonstrates that, in some cases, we can establish quantitative relationships for relatively complex behavior and stimuli.

Although linear modeling provides a powerful technique for establishing quantitative models, even more powerful statistical modeling techniques are now available. Structural equation modeling (SEM) allows psychologists to specify variables and the quantitative relationships among those variables, and then to test the veracity of that model. Although a discussion of SEM is well beyond the scope of this book, let us simply say that it allows researchers to create models, such as those of Wickens, Jensen, or Hunter, and to assign specific numerical relationships among the processes and conditions. These numerical parameters may then be tested statistically and the validity of the hypothesized model tested empirically. The interested reader may consult Raykov and Marcoulides (2006) for an introduction to SEM; however, texts on

Lisrel, AMOS, and EQS—three of the common implementations of SEM—are also widely available. A few examples of the use of SEM can also be found in aviation, including mental workload of student pilots (Sohn and Jo 2003), mental workload and performance in combat aircraft (Svensson et al. 1997), and analysis of the psychometric properties of the U.S. Navy aviation selection test battery (Blower 1998).

All of these models present a view of reality from a different perspective. Arguably, none of them is true in some absolute sense. Rather, they all present a simplified abstraction of reality. Previously, we described the work on modeling of pilot weather-related decision making that we and Mark Wiggins from the University of Western Sydney conducted (Hunter et al. 2003). Although the results of the mathematical modeling process produced very reliable results, we would certainly not argue that humans actually have small calculators in their heads that they use to evaluate such judgments. Rather, we would suggest that whatever is happening during the decision-making process can be predicted rather accurately using our mathematical model.

The distinction is important because later in this book we will discuss many psychological constructs and describe research that shows relationships between a construct and some outcome of interest. For example, we might look at how intelligence relates to completing training successfully. Or, we could examine how a person's internality (the degree to which he or she believes himself or herself in control of his or her destiny) relates to accident involvement. Both intelligence and internality are psychological constructs—convenient titles given to hypothesized underlying psychological traits and capacities. A psychological construct is an abstract theoretical variable invented to explain some phenomenon of interest to scientists. Like most of the topics introduced in this book, the issue of constructs, in particular their measurement, is also the subject of many articles and books in its own right. (For more information, see Campbell and Fiske [1959] regarding the measurement of constructs. Also, for a contemporary problem in construct definition, see the discussion on emotional intelligence by Mayer and Salovey [1993].)

The reader should be aware of the nebulous nature of these constructs and the models that psychologists have devised to describe the relationships among constructs and external events. The insights that the constructs provide and the predictions that may be made from the models are potentially useful, even though the utility of a psychological construct or model is no guarantee of its underlying physical reality.* The reader should not dismiss the constructs as mere "psychobabble" or the models as gross oversimplifications of a complex world. Both can be taken as a means of helping us understand the world by framing it in familiar terms.

Before we leave the discussion of models, constructs, and theories, let us make one last point. Although specifically addressed to the role of Fitts's law, the comments of

---

* This situation is not unique to psychological science. The *Bohr model* of the atom, proposed by Niels Bohr in 1915, is not completely correct, but has many features that are approximately correct and that make it useful for some discussions. At present, the generally accepted theory of the atom is called quantum mechanics; the Bohr model is an approximation to quantum mechanics that has the virtue of being much simpler. Perhaps in 100 years, the quantum mechanical model will be considered a quaint approximation to reality.

Pew and Baron (1983, p. 664) regarding models and theories are broadly applicable. They note:

> There is no useful distinction between models and theories. We assert that there is a continuum along which models vary that has loose verbal analogy and metaphor at one end and closed-form mathematical equations at the other, and that most models lie somewhere in-between. Fitts' law may be placed in this continuum. As a mathematical expression, it emerged from the rigors of probability theory, yet when transplanted into the realm of psychomotor behavior it becomes a metaphor.

1.9   SUMMARY

In this chapter we have tried to introduce the student to some of the concepts and goals of psychology and to delineate some of the domain of aviation psychology. We have also outlined some of the dominant models of human performance that are currently applied to further our understanding of how humans perform in an aviation setting.

Aviation psychology represents an amalgamation of the various approaches and subdisciplines within psychology. In the following chapters we will delve more deeply into the design and development of aviation systems, the selection and training of pilots, and efforts to improve safety from a psychological perspective. Like the rest of the aviation community, aviation psychologists share the ultimate goals of improving safety, efficiency, and comfort. Practitioners of aviation psychology bring to bear the tools and techniques of psychology to describe, predict, understand, and influence the aviation community to achieve those goals.

REFERENCES

Anderson, J. R. 1982. Acquisition of cognitive skill. *Psychological Review* 89:369–406.

Baddeley, A. D. 1986. *Working memory.* Oxford, UK: Oxford University Press.

Blower, D. J. 1998. Psychometric equivalency issues for the APEX system (special report 98-1). Pensacola, FL: Naval Aerospace Medical Research Laboratory.

Bradshaw, B. K. 2003. The SHEL model: Applying aviation human factors to medical error. *Texas D.O.* January: 15–17.

CAA (Civil Aviation Authority). 2002. Safety management systems for commercial air transport operations. CAP 712. Gatwick, UK: Civil Aviation Authority, Safety Regulation Group.

Campbell, D. T., and Fiske, D. W. 1959. Convergent and discriminant validation by the multitrait–multimethod matrix. *Psychological Bulletin* 56:81–105.

CASA (Civil Aviation Safety Authority). 2002. *Safety management systems: Getting started.* Canberra, Australia: Author.

Clarke, R. 1986. *A new approach to training pilots in aeronautical decision making.* Frederick, MD: AOPA Air Safety Foundation.

Corsini, R. J., Craighead, W. E., and Nemeroff, C. B. 2001. Psychology in Germany. In *The Corsini encyclopedia of psychology and behavioral science,* 2nd ed., 633–634. New York: Wiley.

Dar, R. 1987. Another look at Meehl, Lakatos, and the scientific practices of psychologists. *American Psychologist* 42:145–151.

Dreyfus, H., and Dreyfus, S. 1986. *Mind over machine: The power of human intuition and expertise in the era of the computer.* New York: Free Press.

Edwards, E. 1988. Introductory overview. In *Human factors in aviation,* ed. E. L. Weiner and D. C. Nagel, 3–25. San Diego: Academic Press.

Fitts, P. M. 1954. The information capacity of the human motor system in controlling the amplitude of movement. *Journal of Experimental Psychology* 47:381–391. (Reprinted in *Journal of Experimental Psychology: General* 121:262–269, 1992.)

———. 1964. Perceptual skill learning. In *Categories of skill learning,* ed. A.W. Melton, 243–285. New York: Academic Press.

Fitts, P. M., and Posner, M. I. 1967. *Learning and skilled performance in human performance.* Belmont, CA: Brock–Cole.

Foyle, D. C., Hooey, B. L., Byrne, M. D., Corker, K. M., Duetsch, S., Lebiere, C., Leiden, K., and Wickens, C. D. 2005. Human performance models of pilot behavior. In *Proceedings of the Human Factors and Ergonomics Society 49th Annual Meeting,* 1109–1113.

GAIN. 2001. *Operator's flight safety handbook.* Global Aviation Information Network. Chevy Chase, MD: Abacus Technology.

Hawkins, F. H. 1993. *Human factors in flight.* Aldershot, England: Ashgate Publishing Ltd.

Hick, W. E. 1952. On the rate of gain of information. *Quarterly Journal of Experimental Psychology* 4:11–26.

Hunter, D. R. 2002. A proposed human performance model for ADM. Unpublished manuscript. Washington, D.C.: Federal Aviation Administration.

Hunter, D. R., Martinussen, M., and Wiggins, M. 2003. Understanding how pilots make weather-related decisions. *International Journal of Aviation Psychology* 13:73–87.

IACO (International Civil Aviation Organization). 2005. ICAO training page. Retrieved September 15, 2007, from http://www.icao.int/anb/safetymanagement/training%5Ctraining.html

Isaac, A., Shorrock, S. T., Kennedy, R., Kirwan, B., Andersen, H., and Bove, T. 2002. Short report on human performance models and taxonomies of human error in ATM (HERA). Report HRS/HSP-002-REP-02. Brussels: European Organization for the Safety of Air Navigation.

Jensen, R. S. 1988. Creating a "1000 hour" pilot in 300 hours through judgment training. In *Proceedings of the Workshop on Aviation Psychology.* Newcastle, Australia: Institute of Aviation, University of Newcastle.

———. 1995. *Pilot judgment and crew resource management.* Brookfield, VT: Ashgate Publishing.

———. 1997. The boundaries of aviation psychology, human factors, aeronautical decision making, situation awareness, and crew resource management. *International Journal of Aviation Psychology* 7:259–267.

Jensen, R. S., and Benel, R. A. 1977. Judgment evaluation and instruction in civil pilot training (technical report FAA-RD-78-24). Washington, D.C.: Federal Aviation Administration.

Kahneman, D. 1973. *Attention and effort.* Upper Saddle River, NJ: Prentice Hall.

Klayman, J., and Ha, Y. W. 1987. Confirmation, disconfirmation, and information in hypothesis testing. *Psychological Review* 94:211–228.

Kochan, J. A., Jensen, R. S., Chubb, G. P., and Hunter, D. R. 1997. *A new approach to aeronautical decision-making: The expertise method.* Washington, D.C.: Federal Aviation Administration.

Lakatos, I. 1970. Falsification and the methodology of scientific research programs. In *Criticism and the growth of knowledge,* ed. I. Lakatos and A. Musgrave, 91–196. New York: Cambridge University Press.

Landauer, T. K., and Nachbar, D. W. 1985. Selection from alphabetic and numeric menu trees using a touch screen: Breadth, depth, and width. *Proceedings of CHI 85,* 73–78. New York: ACM.

Leiden, K., Laughery, K. R., Keller, J., French, J., Warwick, W., and Wood, S. D. 2001. A review of human performance models for the prediction of human error (technical report). Boulder, CO: Micro Analysis and Design.

Licht, M. H. 2001. Multiple regression and correlation. In *Reading and understanding multivariate statistics,* ed. L. G. Grimm and P. R. Yarnold, 19–64. Washington, D.C.: American Psychological Association.

Mayer, J. D. and Salovey, P. 1993. The intelligence of emotional intelligence. *Intelligence* 17:433–442.

Molloy, G. J., and O'Boyle, C. A. 2005. The SHEL model: A useful tool for analyzing and teaching the contribution of human factors to medical error. *Academy of Medicine* 80:152–155.

Norman, D., and Bobrow, D. 1975. On data-limited and resource-limited processing. *Cognitive Psychology* 7:44–60.

O'Hare, D. 1992. The "artful" decision maker: A framework model for aeronautical decision making. *International Journal of Aviation Psychology* 2:175–191.

Pew, R. W., and Baron, S. 1983. Perspectives on human performance modeling. *Automatica* 19:663–676.

Poletiek, F. H. 1996. Paradoxes of falsification. *Quarterly Journal of Experimental Psychology Section A: Human Experimental Psychology* 49:447–462.

Popper, K. R. 1959. *The logic of scientific discovery.* New York: Harper & Row.

Raykov, T., and Marcoulides, G. A. 2006. *A first course in structural equation modeling.* Mahwah, NJ: Lawrence Erlbaum Associates.

Reason, J. 1990. *Human error.* New York: Cambridge University Press.

Sohn, Y., and Jo, Y. K. 2003. A study on the student pilot's mental workload due to personality types of both instructor and student. *Ergonomics* 46:1566–1577.

Svensson, E., Angelborg-Thanderz, M., Sjöberg, L., and Olsson, S. 1997. Information complexity—Mental workload and performance in combat aircraft. *Ergonomics* 40:362–380.

Transport Canada. 2001. Introduction to safety management systems (TP 13739E). Ottawa: Author

VanLenh, K. 1996. Cognitive skill acquisition. *Annual Review of Psychology* 47:513–539.

Wickens, C. D., and Holland, J. G. 2000. *Engineering psychology and human performance,* 3rd ed. Upper Saddle River, NJ: Prentice Hall.

Wickens, C. D., Goh, J., Hellebert, J., Horrey, W., and Talleur, D. A. 2003. Attentional models of multitask pilot performance using advanced display technology. *Human Factors* 45:360–380.

Wozniak, R. H. 1999. *Classics in psychology, 1855–1914: Historical essays.* Bristol, England: Thoemmes Press.

2 Research Methods and Statistics

2.1 INTRODUCTION

The purpose of research is to gain new knowledge. In order to be able to trust new findings, it is important that scientific methodology is used. The tools that researchers use to conduct research should be described in sufficient detail so that other researchers may conduct and replicate the study. In other words, an important principle is that the findings should be replicable, which means that they are confirmed in new studies and by other researchers. There are several scientific methods available, and the most important aspect is that the methods are well suited for exploring the research question. Sometimes, the best choice is to use an experiment, whereas at other times a survey may be the best choice.

Research ideas may come from many sources. Many researchers work within an area or research field, and a part of their research activity will be to keep updated on unresolved questions and unexplored areas. Other times, the researcher will get ideas from his or her own life or things that happen at work, or the researcher may be asked to explore a specific problem.

Research can be categorized in many ways—for example, basic and applied research. In basic research, the main purpose is to understand or explain a phenomenon without knowing that these findings will be useful for something. In applied research it is easier to see the possibilities for using the research findings for something. Frequently, the boundaries between these two types of research will be unclear, and basic research may later be important as a background for applied research and for the development of products and services. As an example, basic research about how the human brain perceives and processes information may later be important in applied research and in the design of display systems or perhaps for developing tests for pilot selection.

Research should be free and independent. This means that the researcher should be free to choose research methods and to communicate the results without any form of censorship. To what extent the researcher is free to choose the research problem is partly dependent upon where the researcher works; however, frequently one important practical limitation is lack of funding. Even though the researcher may have good ideas for a project and have chosen appropriate methods, the project may not receive any funding.

2.2 THE RESEARCH PROCESS

The research process normally consists of a series of steps (Figure 2.1) that the researcher proceeds through from problem description to final conclusion. The first step is usually a period in which ideas are formulated and the literature on the topic reviewed. The first ideas are then formulated in more detail as problems and hypotheses. Some may be very descriptive—for example, the prevalence of fear of flying in the population. At other times, the purpose may be to determine the cause of something—for example, whether a specific course aimed at reducing fear of flying is in fact effective in doing so.

The next step will be to choose a method well suited for studying the research problem. Sometimes aspects other than the nature of the problem or hypothesis will influence the choice of method—for example, practical considerations, tradition, and ethical problems. Within certain disciplines, some methods are more popular than others, and choice of methods may also depend on the training that the researcher has received. In other words, there will likely be many aspects involved when choosing research methods and design in addition to the nature of the research question. The next step in the research process involves data collection. Data may be quantitative, implying that something is measured or counted that may be processed using statistical methods, or that data may be qualitative, often involving words or text. The majority of research conducted within aviation is based on quantitative data, and this chapter will focus on how to collect, process, and interpret such data.

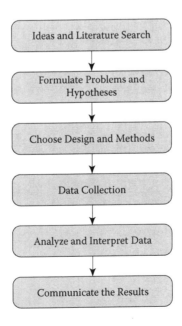

FIGURE 2.1 The research process.

2.3   LITERATURE REVIEW AND RESEARCH QUESTIONS

There are a large number of scientific journals in psychology, and some of them publish aviation-related research. One example is the *International Journal of Aviation Psychology,* which is an American journal published by Taylor & Francis. Another journal published by the U.S. Federal Aviation Administration (FAA) is the *International Journal of Applied Aviation Studies,* which is available online.* In addition, a medical journal called *Aviation, Space, and Environmental Medicine* is published by the Aerospace Medical Association. A number of other psychology journals will publish articles in aviation psychology and other related areas of work and organizational psychology—for example, *Human Factors, Journal of Applied Psychology,* and *Military Psychology.* Articles from these journals may be accessed through a library. One can also visit the journal's Web site, but it may be necessary to purchase the article, at least the more recent issues. In addition, the FAA has produced a large number of reports over the years that are available online, free of charge. A general literature search using a regular search engine may also result in relevant articles and reports, but it is important to evaluate the source and quality of the information critically.

Before they are published, scientific articles have been through a peer-review system. This means that two or three other researchers in the field will review the article. The author then receives the feedback from the reviewers and the editor, and in most cases the article will have to be revised before it is published. Sometimes, the quality of the article is too poor compared to the standards of the journal, and the author is not given the option to revise and resubmit. The journals vary in relation to the proportion of submitted articles that are accepted and how often articles from the journal are cited by other authors.

Research in aviation psychology is also presented at conferences. One of the organizers of such conferences is EAAP (European Association of Aviation Psychology). EAAP is an association of aviation psychologists now more that 50 years old that organizes conferences every 2 years. In addition, the International Symposium in Aviation Psychology is organized every 2 years and is usually held in Dayton, Ohio. Approximately every 3 years, the Australian Association for Aviation Psychology organizes a conference in Sydney. Proceedings are usually published after the conferences, and they include the papers presented at the conference, usually in the form of short articles. These are available for those attending the conference and sometimes also through the Web sites of the organizers. Researchers† often present their results at a conference before the results appear in a journal, so attending conferences may provide a snapshot of the latest news in the area.

---

* Web links for this and other sources are available in the listings given in Chapter 9.
† Researchers are usually thrilled to get a request for their publications, so if an article cannot be found through a library or on the Web, try sending a note to the author and asking for a copy. Many of the researchers in this area are members of the Human Factors and Ergonomics Society, and their e-mail addresses can be found through that organization.

2.4 RESEARCH PROBLEMS

Not all questions can be answered through research, and even if the questions or problems can be examined by scientific methods, it is often necessary to restrict the study to some selected questions. One type of research problem can be categorized as *descriptive*. This term is used when the purpose is to describe a phenomenon or how things co-vary. An example of a descriptive research problem would be to assess the level of stress among air traffic controllers or to examine whether a test can be used to select cabin crew members. Sometimes, in addition to an overall research problem, more specific questions may be formulated as hypotheses. Suppose a survey on job stress is conducted among air traffic controllers. The researcher might want to examine the hypothesis that there is an association between lack of social support and experienced stress. Such hypotheses are usually based on a theory or on earlier research findings.

If a correlation between two variables—for example, between social support and stress—is discovered in a survey, it does not necessarily mean that a causal relationship exists between the two variables. It may be a causal link, but a correlation is not sufficient to determine this. In addition, we need to know that social support precedes the feeling of stress, and preferably, we should know something about the mechanism behind or how social support reduces or acts as a buffer to stress. Perhaps a third variable affects both variables studied and is the real cause of variations in both social support and stress. If a causal relationship is the main focus of the study, then the research question will have to be formulated differently than in descriptive research, and it will also require a different research design from that used in descriptive research. The best way to examine causal problems is usually by means of an experiment.

2.5 VARIABLES

Variables are aspects or attributes of a person or phenomenon studied; they may have different values depending on what is being measured. Some examples include the age of a person or the number of hours of flying experience. Alternatively, we may be interested in personal characteristics such as personality traits—for example, extroversion or whether the person has completed a CRM (crew resource management) course. Variables manipulated by the researcher are usually labeled independent variables (e.g., whether a person receives a course or not). The variables studied afterward (e.g., improved communication skills or fewer operational mistakes) are called dependent variables.

Variables also differ in how they are measured or assessed. Some variables are easy to assess; for example, age could be measured in years (at least for adults) or salary measured in dollars or Euros. Both age and salary are continuous variables measured on a scale with equal intervals between the values. Many psychological variables are measured using a five-point scale or a seven-point scale where the person is asked to indicate how much she or he agrees with a statement. Sometimes fewer categories are used, and it can be argued whether such scales can be seen as continuous or not; this has implications for the statistical methods that may be used for analyzing the data.

Some variables are categorical; for example, gender has two categories (man/woman), whereas masculinity could be measured on a continuous scale. Numbers may also be used for categorical variables, but would only say something about group membership (e.g., men are given the value 1 and women the value 2) and simply serve as labels. The number would only indicate that they are different, and the letters A and B could be used instead. Variables may, in other words, have many values and the numbers would contain different types of information depending on the type of variable and what is being measured.

2.6 DESCRIPTIVE METHODS AND MEASUREMENT

Research usually requires measurement or some form of categorization. Some variables are easy to assess, and others are not. Many psychological constructs are not easily observed—for example, intelligence, anxiety, or personality traits. These constructs will have to be operationalized before they can be measured. Psychological tests are one way of measuring these constructs, in addition to interviews and observations.

2.6.1 PSYCHOLOGICAL TESTS

A psychological test is a standardized procedure for assessing the amount or magnitude of something. It could involve a handful of questions or an extensive procedure involving equipment and computers. Psychological tests may be used for many different purposes, including clinical use, personnel selection, and research. Regardless of purpose, it is important that the tests be of high quality, which usually involves three requirements: reliability, validity, and appropriate norms. Each of these concepts will be discussed in detail on the following pages.

There are many requirements for tests and test users. The European psychological associations have agreed on a common set of guidelines (International Guidelines for Test Use), which can be found on the Web sites of the national psychological associations or the International Test Commission. The American Psychological Association (APA 1999) has for many years published a book called *The Standards for Educational and Psychological Testing,* which explains reliability and validity in addition to providing guidelines for how tests should be used appropriately, professionally, and in an ethical manner.

2.6.2 CLASSICAL TEST THEORY

In psychology, we usually assume that a person's test score consists of two components that together constitute what is called the observed score. One component is the person's true score, while the other is the error term. This can be expressed as follows: Observed score = true score + error part. If the same person is tested several times under the same conditions, we would expect a similar but not identical test score every time. The observed scores will vary slightly from time to time. If the error term is small, then the variations will be smaller than if the error term is large. In addition, we make the assumption that the error is unsystematic—for example, that the error does not depend on a person's true score. This model of measurement is called *classical test theory* (Magnusson 2003) and is the starting point for many of the psychological tests used today.

A more recent development is *item response theory* (Embretson and Reise 2000), which represents a different way of studying test scores. It is the starting point of so-called adaptive testing, in which the degree of difficulty on tasks is tailored to the individual level. The theory is based on the assumption that the probability of answering correctly is a function of a latent trait or ability. Item response theory makes stronger assumptions than classical test theory, but the advantage is a more detailed analysis of the items. One disadvantage is the lack of user-friendly software to conduct the analyses compared to the analyses performed in classical test theory, which may be performed with most statistical programs. In the years to come, more tests will certainly be developed based on this theory.

2.6.3 RELIABILITY

Within classical test theory, reliability is defined as the correlation between the two parallel tests—that is, two similar but not identical tests. The correlation coefficient is a statistical measure of covariation between two variables. This index will be further discussed in Section 2.8 on statistics. In other words, if we test a group of people twice using two parallel tests, a large correlation is expected between the test scores if the error term is small. However, it is often difficult and time consuming to create parallel tests, so a different approach to examining the reliability of the test scores is needed.

One way to estimate reliability is to test the same group of people twice with the same test (e.g., after a few months). This is called *test–retest reliability.* Another common approach is to divide the test into two parts—for example, by taking every other question in each section. Then a group of people is tested and two scores are calculated for each person, one for each part of the test. Finally, the correlation is calculated between the two parts. However, this approach will result in an estimate of the reliability of a test with only 50% of the items or half the length of the original test. It is possible to correct the reliability estimate for this using a formula that provides an estimate of the reliability for the entire test.

Of course, we can divide the test into two parts in many ways, depending on how the questions or items are split. One form of reliability that is frequently used is the Cronbach's alpha, which is a kind of average split-half reliability of all the possible split-half reliability estimates for a given measure. The various types of reliability will provide different types of information about the test scores. Test–retest reliability will say something about the stability over time, and the other forms of reliability will provide information about the internal consistency (split-half and Cronbach's alpha). The calculated correlations should be as high as possible, preferably .70 or .80, but sometimes lower values may be accepted. One factor affecting test reliability is the number of questions: The more questions there are, the higher the reliability is. In addition, it is important that the test conditions and scoring procedures be standardized, which means that clear and well-defined procedures are used for all the subjects.

2.6.4 Validity

The most important form of test validity is *construct validity*. This refers to the extent to which the test measures what it purports to be measuring. Does an intelligence test measure intelligence or is the test only a measure of academic performance? There is no simple solution to the problem of documenting adequate construct validity. Many strategies may be used to substantiate that the test actually measures what we want it to measure. If, for example, we have developed a new intelligence test, then one way to examine construct validity would be to investigate the relationship between the new test and other well-established intelligence tests.

Another form of validity is *criterion-related validity*. This refers to the extent to which the test predicts a criterion. If the criterion is measured about the same time as the test is administered, then the term "concurrent validity" is used, in contrast to "predictive validity," which is used when a certain time period has passed between testing and measurement of the criterion. Predictive validity asks the question: Can the test scores be used for predicting future work performance? The predictive validity is usually examined by calculating the correlation between the test scores and a measure of performance (criterion).

Content validity, which is the third type of validity, concerns the extent to which the test items or questions are covering the relevant domain to be tested. Do the exam questions cover the area to be examined or are some parts left out?

These three forms of validity may seem quite different, and the system has received some criticism (Guion 1981; Messick 1995). An important objection has been that it is not the test itself that is valid; rather, the conclusions that we draw on the basis of test scores must be valid. Messick (1995) has long argued that the examination of test validity should be expanded to include value implications inherent in the interpretation of a test score as well as social consequences of testing. This would involve an evaluation to see whether the use of a particular test may have unfortunate consequences—for example, that special groups are not selected in connection with the selection of a given job or education.

2.6.5 Test Norms and Cultural Adaptation

In addition to reliability and validity, it is often desirable that the test be standardized. This means that the test scores for a large sample of subjects are known so that a person may be compared with the mean of these scores. Sometimes it may be appropriate to use a random sample of the population when establishing test norms, whereas at other times more specialized groups are more relevant. Imagine a situation where a person is tested using an intelligence test and the result is calculated as the number of correct responses. Unless we compare the result with something, it is hard to know whether the person performed well or not. The result could be compared to the average number of correct responses based on other adults in the same age group and from the same country. The tests should be administered in the same way for all who have been tested. Everyone receives the questions in the same order, with the same instructions, and with specific scoring procedures.

This is an important principle in order to be able to compare performance. If a test is not developed and norms established in the country where the test will be used, then it is necessary to translate and adapt the test to the new conditions. For example, when a test developed in the United States is used in Norway, the test needs to be translated to Norwegian. A common procedure is first to translate the test into Norwegian and then for another bilingual person to translate the work back to the original language. This procedure may reveal problems in the translation that should be resolved before the test is used. Even if we can come up with a good word-for-word translation, test reliability and validity should be examined again, and national norms should be established.

2.6.6 QUESTIONNAIRES

A questionnaire is an efficient way to collect data from large groups. It is also easier to ensure that the person feels that he or she can respond anonymously (e.g., compared to an interview). A questionnaire may consist of several parts or sections (e.g., one section on background information such as age, gender, education, and experience). The questionnaire will include specific questions designed for the purpose and may also include more established scales that, for example, measure personality characteristics such as extroversion. The answers can be open ended or with closed options. Sometimes a five-point or seven-point scale is used where people can indicate their opinion by marking one of the options. If the questionnaire involves using established scales—for example, to measure satisfaction in the workplace—it is important to keep the original wording and the response options of these scales identical to the original measurement instrument. Changes may alter the psychometric properties of the instrument and make comparisons with other studies difficult.

To formulate good questions is an art, and often there will be a lot of work behind a good questionnaire. It is important to avoid formulating leading or ambiguous questions, to use a simple language, and avoid professional terminology and expressions. A questionnaire that looks appealing and includes clear questions increases the response rate. The length of the questionnaire is also related to the response rate, so shorter questionnaires are preferred; thus, designers must think carefully about whether all the questions really are necessary. A good summary of advice when formulating questions and conducting a survey is provided by Fink and Kosecoff (1985).

Researchers would like the response rate (the proportion of people who actually complete the survey and/or send back the questionnaire) to be as high as possible. Some methods books claim that it should be at least 70%, but this proves difficult to achieve in practice, even after a reminder has been sent to all the respondents. A meta-analysis of studies in clinical and counseling psychology summarized 308 surveys and found an average response rate of 49.6% (Van Horn, Green, and Martinussen 2009). The survey showed that the response rate increased by an average of 6% after the first reminder. Response rate also declined over the 20-year period covered by the meta-analysis (1985–2005).

2.6.7 INTERNET

Many surveys are conducted via the Internet, and various programs can be used to create Internet-based questionnaires. Some of the programs must be purchased, but other applications can be freely downloaded via the Web. One example of free software is Modsurvey, which was developed by Joel Palmius in Sweden (http://www. modsurvey.org/). Participants in Internet surveys may be recruited by sending them an e-mail or by making the address of the Web site known to the audience in many ways. It is often difficult to determine what the response rate is in online surveys. This is due in part to the fact that e-mail addresses change more frequently than residential addresses, and thus it is difficult to know how many people actually received the invitation. In cases where participants are recruited through other channels, it may also be difficult to determine how many people were actually informed about the survey.

Internet surveys are becoming very popular because they are efficient and save money on printing, postage, and also punching of data. However, these surveys may not be the best way to collect data for all topics and all participant groups. Not everyone has access to a PC, and not all people will feel comfortable using it for such purposes.

2.6.8 INTERVIEW

An interview can be used for personnel selection and as a data collection method. The interview may be more or less structured in advance; that is, the extent to which the questions are formulated and the order can be determined in advance. When exploring new areas or topics, it is probably best for the questions to be reasonably open; at other times, specific questions should be formulated and, in some instances, both the questions and answering options will be given. If the questions are clearly formulated in advance, it will probably be sufficient for the interviewer to write down the answers. During extensive interviews, it may be necessary to use a tape recorder, and the interview will have to be transcribed later. An interview is obviously more time consuming to process than a questionnaire, but probably more useful when complex issues have to be addressed or new themes explored. It may therefore be wise to conduct some interviews with good informants before developing a questionnaire.

Before the interview begins, an interview guide with all the questions is usually constructed, and if multiple interviewers are used, they should all receive the necessary training so that the interviews are conducted in the same way. It is also important that interviewers are aware of the possible sources of error in the interview and how the interviewer may influence the informants with his or her behavior.

2.6.9 OBSERVATION

Like the interview, observation of people may be more or less structured. In a structured observation, the behavior to be observed is specified in advance, and there are clear rules for what should be recorded. An example would be an instructor who is evaluating pilots' performance in a simulator. Then different categories should be specified and what constitutes good and poor performance should be outlined in advance. The observers may be a part of the situation, and the observed person may not even be aware that he or she is observed. This is called hidden observation.

An obvious problem with observation is that people who are observed may be influenced by the fact that an observer is present. A classic experiment that illustrates this is the Hawthorne study, where workers at a U.S. factory producing telephone equipment were studied. The purpose was to see whether various changes in lighting, rest hours, and other working conditions increased the workers' production. Irrespective of the changes implemented—increased lighting or less light—production increased. One interpretation of these findings was that being observed and receiving attention affected individuals' job performances. The study is described in most introductory textbooks in psychology. The fact that people change behavior when being observed has subsequently been given the term "Hawthorne effect" after the factory where these original studies were conducted in the 1920s.

The study has since been criticized because the researchers failed to consider a number of other factors specific to the workers who participated in the study. One difference was that women who took part in the study gained feedback on their performance and received economic rewards compared to the rest of the factory workers (Parsons 1992). This shows that studies may be subject to renewed scrutiny and interpretation more than 60 years after they have been conducted. Regardless of what actually happened in the Hawthorne plant, it is likely that people are influenced by the fact that they are observed. One way to prevent this problem is to conduct a so-called hidden observation. This is not ethically unproblematic, especially if one is participating in the group observed. The situation is different if large groups are observed in public places and the individuals cannot be identified. For example, if the researcher is interested in how people behave in a security check, this may be an effective and ethical research method.

Often, in the beginning of a study, those who are observed are probably aware of the fact that an additional person is present in the situation. After some time, however, the influence is probably less as the participants get used to having someone there and get busy with work tasks to be performed.

2.7   EXPERIMENTS, QUASI-EXPERIMENTS, AND CORRELATION RESEARCH

In a research project, it is important to have a plan for how the research should be conducted and how the data should be collected. This is sometimes called the design of the study; we will describe three types of design.

2.7.1   EXPERIMENTS

Many people probably picture an experiment as something that takes place in a laboratory with people in white coats. This is not always the case, and the logic behind the experiment is more crucial than the location. An important feature of an experiment is the presence of a control group. The term "control group" is used for one of the groups that receives no treatment or intervention. The control group is then compared to the treatment group, and the differences between the groups can then be attributed to the treatment that one group received and the other did not. This requires that the groups be similar, and the best way to ensure this is by randomly assigning subjects to the two conditions.

Another important feature of an experiment is that the researcher can control the conditions to which people are exposed. Sometimes, several experimental groups receive different interventions. For example, a study may include two types of training, both of which will be compared to the control group that receives no training. Alternatively, there may be different levels of the independent variable; for example, a short course may be compared to a longer course.

2.7.2 QUASI-EXPERIMENTS

A compromise between a true experiment and correlational studies is a so-called quasi-experimental design. The purpose of this design is to mimic the experiment in as many ways as possible. An example of a quasi-experiment is to use a comparison group and treatment group to which the participants have not been randomly assigned. It is not always possible to allocate people at random to experimental and control conditions. Perhaps those who sign up first for the study will have to be included in the experimental group, and those who sign up later will have to be included in the comparison group. Then the researcher has to consider the possibility that these groups are not similar.

One way to explore this would be to do some pretesting to determine whether these groups are more or less similar in relation to important variables. If the groups differ, this may make it difficult to draw firm conclusions about the effect of the intervention. There are other quasi-experimental designs, such as a design without any control or comparison group. One example would be a pretest/posttest design where the same people are examined before and after the intervention. The problems associated with this design will be addressed in the section on validity.

2.7.3 CORRELATIONAL RESEARCH

It is not always possible to conduct a real experiment for both practical and ethical reasons. For example, it may not be possible to design an experiment in which the amount of social support employees receive from the leader is manipulated. This approach will be viewed as unethical by most people, but studying natural variation in this phenomenon is possible. Research in which working conditions are studied will often include a correlational design. The purpose may be to map out various work demands such as workload and burnout. After these variables have been studied, various statistical techniques may be used to study the relationship between these variables. Also, more complex models of how multiple variables are connected with burnout can be studied, in addition to examining the extent to which burnout can be predicted from work-related factors and personal characteristics.

2.8   STATISTICS

An important part of the research process occurs when the results are processed. If the study includes only a few subjects, then it is easy to get an overview of the findings. However, in most studies, a large number of participants and variables are included, so a tool is clearly needed to get an overview. Suppose that we have sent out a questionnaire to 1,000 cabin crew members to map out what they think about their working environment. Without statistical techniques, it would be almost impossible to describe the opinions of these workers.

Statistics can help us with three things:

- sampling (how people are chosen and how many people are required);
- describing the data (graphical, variation, and the most typical response); and
- drawing conclusions about parameters in the population.

In most cases, we do not have the opportunity to study the entire population (e.g., all pilots and all passengers), and a smaller group needs to be sampled. Most statistical methods and procedures assume a random selection of subjects, which means that all participants initially have equal chances to be selected. There are also other ways to sample subjects; for example, stratified samples may be employed where the population is divided into strata and then subjects are randomly selected from each stratum. These sampling methods are primarily used in surveys where one is interested in investigating, for example, how many people sympathize with a political party or the extent of positive attitudes toward environmental issues. An application of statistics is thus to determine how the sampling should be done and, not least, how many people are needed in the study.

After the data are collected, the next step is to describe the results. There are many possibilities, depending on the problem and what types of data have been collected. The results may be presented in terms of percentages, rates, means, or perhaps a measure of association (correlation). Graphs or figures for summarizing the data could also be used.

The third and last step is the deduction from the sample to the population. Researchers are usually not satisfied with just describing the specific sample, but rather want to draw conclusions about the entire population. Conclusions about the population are based on findings observed in a sample. One way to do this is through hypothesis testing. This means that a hypothesis about the population is formulated that can be tested using results from the sample.

The second procedure involves estimating the results in the population on the basis of the results in the sample. Suppose we are interested in knowing the proportion of people with fear of flying. After conducting a study measuring this attitude about flying, we will have a concrete number: the proportion of our subjects who said they were afraid of flying. Lacking any better estimates, it would then be reasonable to suggest that, among the population at large, approximately the same proportion as we have observed in the sample has a fear of flying. In addition, we might propose an interval that is likely to capture the true proportion of people with fear of flying in the population. These intervals are called confidence intervals. If many people are included in the sample, then these intervals are smaller; that is, the more people who are in our sample, the more accurate is our estimate.

2.8.1 DESCRIPTIVE STATISTICS

The most common measure of central tendency is the arithmetic mean. This is commonly used when something is measured on a continuous scale. The arithmetic mean is usually denoted with the letter M (for "mean") or $\overline{X}$. An alternative is the median, which is the value in the middle, after all the values have been ranked from lowest to highest. This is a good measure if the distribution is skewed—for example, if the results include some very high or very low values. The mode is a third indicator of central tendency, which simply is the value with the highest frequency. Thus, it is not necessary that the variable be continuous to use this measure of central tendency.

In addition to a measure of the most typical value in the distribution, it is also important to have a measure of variation. If we have calculated the arithmetic mean, it is common to use the standard deviation as a measure of variation. Formulated a little imprecisely, the standard deviation is the average deviation from the mean. If results are normally distributed—that is, bell shaped—and we inspect the distribution and move one standard deviation above and one standard deviation below the average, then about two-thirds of the observations fall within that range. Including two standard deviations on both sides of the mean, then about 95% of the observations will be included. The formulas for calculating the arithmetic mean and standard deviation are

$$\overline{X} = \frac{\sum_{i=1}^{N} X_i}{N}$$

$$SD = \sqrt{\frac{\sum_{i=1}^{N} (X_i - \overline{X})^2}{N-1}}$$

where
 N = sample size
 X_i = test score
 $\overline{X}$ = mean test score

As an example, we have created a simulated data set for 10 people (Table 2.1). Assume that people have answered a question in which they are asked to specify on a scale from 1 to 5 the degree of fear of flying. In this case, the higher the score is, the greater is the discomfort. In addition, age and gender are recorded. Remember that these data have been fabricated. Suppose we are interested in describing the group in terms of demographic variables and the level of fear of flying. Both age

TABLE 2.1
Constructed Data Set

Person	Fear of Flying	Age	Sex
1	5	30	Female
2	4	22	Female
3	4	28	Male
4	3	19	Male
5	3	20	Female
6	2	21	Female
7	2	22	Male
8	1	24	Male
9	1	23	Male
10	1	21	Male

and fear can be said to be continuous variables, and then the arithmetic mean and standard deviation are appropriate measures of central tendency and variation. If a statistical program is used to analyze the data, each person must be represented as a line in the program. The columns represent the different variables in the study. If a widely used statistics program in the social sciences and medicine (Statistical Package for the Social Sciences/SPSS) is applied, the output would look like the one in Table 2.2.

The figures in italics are the computed means and standard deviations for fear of flying and age, respectively. In addition, the minimum and maximum scores for each of the variables are presented. For the variable gender, it is not appropriate to

TABLE 2.2
SPSS Output

Descriptive Statistics

	N	Range	Min.	Max.	Sum	Mean	Std. Deviation
Fear of flying	10	4.00	1.00	5.00	26.00	*2.6000*	*1.42984*
Age	10	11.00	19.00	30.00	230.00	*23.0000*	*3.49603*
Valid N (listwise)	10						

Sex

		Frequency	%	Valid %	Cumulative %
Valid	F	4	40.0	*40.0*	40.0
	M	6	60.0	*60.0*	100.0
	Total	10	100.0	*100.0*	

calculate the mean score, and the best way to describe gender would be to specify the number of women and men included in the study. If the sample is large, it would be wise to give this as a percentage.

The researcher may also be interested in exploring the relationship between age and fear of flying. Because both variables are continuous, the Pearson product-moment correlation may be used as an index of association. The correlation coefficient indicates the strength of the relationship between two variables. It is a standardized measure that varies between −1.0 and +1.0. The number indicates the strength of the relationship and the sign indicates the direction. A correlation of 0 means that there is no association between the variables, while a correlation of −1.0 or 1.0 indicates a perfect correlation between variables in which all points form a straight line. Most correlations that we observe between two psychological variables will be considerably lower than 1.0. (See Table 2.4 for a description of what can be labeled a small, medium, or high correlation.)

Whenever a correlation is positive, it means that high values on one variable are associated with high values on the other variable. A negative correlation means that high values on one variable are associated with low values for the other variable. Correlation actually describes the extent to which the data approximate a straight line, and this means that if there are curvilinear relationships, the correlation coefficient is not an appropriate index. It is therefore wise to plot the data set before performing the calculations.

The formula for calculating the product-moment correlation is

$$r = \frac{\sum_{i=1}^{N}(X_i - \bar{X})(Y_i - \bar{Y})}{(N-1)S_x S_y}$$

where
S_x, S_y are the standard deviations of the two variables
$\bar{X}$ and $\bar{Y}$ are mean scores
N is sample size

In the example in Figure 2.2, the plot indicates a tendency for higher age to be associated with a higher score in fear of flying (i.e., a positive correlation). According to the output, the observed correlation is .51, which is a strong correlation between the variables. In addition, a significance test of the correlation is performed and reported in the output, but we will return to this later.

Sometimes we want to examine in more detail the nature of the relationship between the two variables beyond the strength of the relationship. For example, we may want to know how much fear of flying would increase when increasing the age by 1 year (or perhaps 10 years). Using a method called regression analysis, we can calculate an equation for the relationship between age and fear of flying that specifies the nature of the relationship. Such a regression equation may have one or more independent variables. The purpose of the regression analysis is to

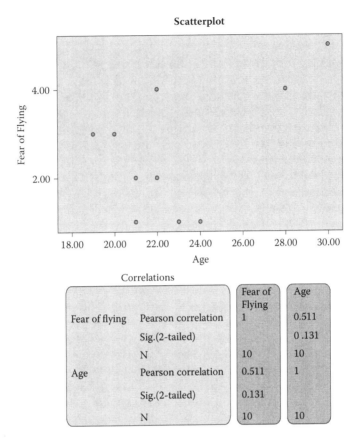

FIGURE 2.2   Correlation: scatter plot and calculations.

predict as much as possible of the variation in the dependent variable, in this case fear of flying.

The correlation coefficient is the starting point for a series of analyses—among other things, factor analysis. The purpose of factor analysis is to find a smaller number of factors that explain the pattern of correlations among many variables. Suppose we have tested a group of people with a range of ability tests that are then correlated with each other. Some of the tests will be highly correlated with each other while others will be less correlated. Using factor analysis, it is possible to arrive at a smaller number of factors than the number of tests that will explain the pattern in the correlation matrix. Perhaps the tests can be grouped into two groups: one including tests that measure mathematical abilities and one consisting of tests that measure verbal abilities. The results of the factor analysis may be the identification of two underlying factors that might be named after tests that are intertwined with each of them. Here one could perhaps say that the tests measure the two abilities—namely, linguistic and mathematical skills.

There are two main strategies when performing a factor analysis: exploratory factor analysis and confirmatory factor analysis. In the first case, a factor structure is suggested by the statistical program based on the specific intercorrelations in the data set. In confirmatory factor analysis, the researcher suggests the number of factors and the structure among these. The model is then compared to the actual data and the correspondence between model and data is examined.

Returning to our data set, we may want to determine whether any gender differences exist. Based on the SPSS output in Table 2.3, we see that women score higher on fear of flying compared to men in this group, although the standard deviations are very similar. Sometimes it can be difficult to evaluate whether such a difference can be said to be big or small. This obviously depends on how well one knows the scale used. An alternative is to transform the difference between the means to a more familiar scale—for example, in the form of a standard deviation. These standardized scores are called effect sizes (ES) and are calculated as $ES = (\bar{X} - \bar{Y})/SD_{pooled}$.

In this calculation, the difference between means is divided by a pooled standard deviation. The standard deviations in this example are approximately equal for the groups and we can use $SD_{pooled} = 1.3$. ES is then $(3.5 - 2.0)/1.3 = 1.15$. This can be described as a large difference, and it means that the difference between women and men is slightly more than one standard deviation. In addition, the confidence interval for the difference is large, ranging from −0.398 to 3.398. The confidence interval is

TABLE 2.3
SPSS Output: t-Test of Differences between Means

Group Statistics

	Sex	N	Mean	Std. Deviation	Std. Error Mean
Fear of flying	Female	4	3.5000	1.29099	.64550
	Male	6	2.0000	1.26491	.51640

Independent Samples Test

	Levene's Test for Equality of Variances		t-Test for Equality of Means					95% Confidence Interval of the Difference	
Fear of Flying	F	Sig.	t	df	Sig. (two-tailed)	Mean Difference	Std. Error Difference	Lower	Upper
Equal variances assumed	.000	1.000	1.823	8	.106	1.50000	.82285	−.39750	3.39750
Equal variances not assumed			1.815	6.477	.116	1.50000	.82664	−.48715	3.48715

estimated to capture the actual difference between women and men in the population with a high certainty.

The range also includes zero, which means that we cannot rule out the possibility that the difference between women and men in the population is zero.

2.8.2 Inferential Statistics

We have now described the variables and also the associations between some of these variables for the sample. The next step would be to draw conclusions about the population from which these individuals were randomly picked. Several interesting hypotheses can be tested. For one, we found a positive correlation between age and fear of flying. The question is: Will there also be a correlation in the population?

This question is restated in the form of a null hypothesis (H_0). The null hypothesis in this example is that the correlation is zero in the population. The alternative hypothesis (H_1) would be that the correlation is greater than zero or less than zero. It is now possible to calculate a number (a test statistic) that will help us decide whether our current results support the null hypothesis or the alternative hypothesis.

In Figure 2.2, the results from this significance test are presented. The number in Figure 2.2 indicates the probability of observing the correlation we have obtained, or a stronger correlation when the true correlation in the population is zero (i.e., H_0 is true). The probability in this case is .13, and this is higher than what we usually would accept. As a rule, the calculated probability should be lower than .05 and preferably .01. This limit is chosen in advance and is called the significance level. By choosing a given significance level, we decide on how "strict" we will be. To put it in a slightly different way: How much evidence is required before the null hypothesis can be discarded? In this example, we cannot reject the null hypothesis because .13 > .05. It is important to remember that the sample in this case is very small (only 10 people); for most purposes, this is too small a sample when conducting a study.

If we want to test whether the difference between women and men is significant, we also need to formulate a null hypothesis. The null hypothesis in this case will be that the difference between the mean scores for women and men is zero in the population (H_0: mean score for women in population–mean score for men in the population = 0). This is another way of saying that the two means are equal. In addition, we need an alternative hypothesis if the null hypothesis is rejected. This hypothesis would be that the difference between the means is different from zero.

Sometimes we choose a specific direction for the alternative hypothesis based on previous findings or theory. Suppose previous studies have shown that women score higher than males on measures of anxiety and depression. We could then formulate an alternative hypothesis that indicates that the difference is greater than zero (H_1: mean score for women in population–mean score for men in the population > 0). In this case, a so-called one-tailed significance test could be performed; if we do not have a specific hypothesis in advance, both options would have to be specified, and this is called a two-tailed test. This represents a more conservative approach and is therefore used more frequently. Most statistics programs therefore use a two-tailed test as the default option.

An example of a significance test is presented in Table 2.3. In the top part of the table, the mean scores for men and women are presented. For women, it is 3.5; men score lower and achieve 2.0. Is this difference large enough for the result to be significant? In Table 2.3, a t-value ($t = 1.82$) and the corresponding probability (.106) are presented. The calculations are based on the assumption that the difference really is zero in the population (H_0 is true). This is a higher probability than is commonly used as a limit (e.g., .05), and thus the null hypothesis cannot be rejected. A prerequisite for such tests, at least if the samples are small, is that the variations in the two groups are equal. The table also shows a test that can be used if this assumption does not hold.

There are a number of different significance tests depending on the hypotheses and how data are collected. In our example, the appropriate test is called a t-test for independent data because two independent groups were examined. If a different design had been used and, for example, the same person had been examined twice, a slightly different version of this significance test is appropriate (dependent samples t-test). When several groups are investigated, there will be many group means. Comparing multiple group means can be done with analysis of variance (ANOVA). The basic principles involved in the significance testing, however, are the same:

- Formulate H_0 and H_1.
- Select the significance level (.01 or .05).
- Perform calculations. What is the probability of getting the results if H_0 is true?
- Conclude. Can H_0 be rejected?

2.8.3 TYPE 1 AND TYPE 2 ERRORS

When a significance test is conducted, several decisions can be made—two correct and two incorrect. Correct decisions are to reject the null hypothesis when it is wrong and keep it when it is correct. Wrong decisions are to retain the null hypothesis when it is incorrect (type 2 error) or reject it when it is correct (type 1 error). The probability of making a type 1 error is set by the significance level. If the researcher wants to be very confident that such a mistake is avoided, a more stringent significance level should be chosen. To avoid type 2 error, it is important that the study be conducted with a sufficient number of people so that true correlations or differences between groups are discovered.

Somewhat imprecisely, one can say that significance is a function of both the effect size and the sample size. This means that if we study a very strong association (high correlation), a small sample may be sufficient. If the relationship is weaker (a smaller correlation), a much larger sample is needed to detect this. It can sometimes be difficult to determine the necessary sample before the survey is conducted if the size of the effect is unknown. If similar studies have already been published, these results can be used to calculate the sample size needed. In Table 2.4, some calculations have been performed for small, medium, and large effects in order to estimate the sample sizes needed. It is, of course, desirable that the study should have sufficient statistical power. This means that the probability should be high (e.g., .80) for correctly rejecting the null hypothesis.

TABLE 2.4
Required Sample Size

	Effect	N (One-Tailed Test)	N (Two-Tailed Test)
	Correlation		
Small effect	$r = .10$	600	770
Medium effect	$r = .30$	60	81
Large effect	$r = .50$	20	25
	Differences Between Two Groups		
Small effect	ES = .20	310 (in each group)	390 (in each group)
Medium effect	ES = .50	50 (in each group)	63 (in each group)
Large effect	ES = .80	20 (in each group)	25 (in each group)

Notes: Significance level = .05; statistical power = .80.

Different programs can be used to perform such calculations; the calculations in Table 2.4 are performed with a program called "Power and Precision" (Borenstein, Rothstein, and Cohen 2001). When the calculations are performed, it is necessary to specify the size of the effect, significance level (one-tailed or two-tailed test), and the desired statistical power.

2.9   DESIGN AND VALIDITY

Every research design will have advantages and disadvantages, and the researcher needs to consider every option and possible methodological problem carefully before choosing the study design. In addition, practical, ethical, and economic factors will have to be considered. An important aspect when choosing design is the impact this will have on validity. A very short and simple explanation of validity is that it relates to how well we can trust the conclusions that are drawn from a study.

There are different forms of validity related to study design, such as *statistical validity, internal validity, external validity,* and *construct validity.* In short, statistical validity is related to obtaining a significant effect, internal validity is about causality, construct validity has to do with measuring constructs, and external validity has to do with generalizing findings over time, places, and people. This validity system was first introduced by Cook and Campbell in 1979; a revised edition of their classic book was published in 2002 (Shadish, Cook, and Campbell 2003). The system is particularly intended for research addressing causal problems, although some types of validity may be relevant to more descriptive research.

2.9.1   Statistical Validity

This type of validity relates to the extent to which we can trust the statistical conclusions from a study. Is there a significant difference between the two groups studied (experimental and control groups), and is the difference of a certain magnitude? To investigate this, we usually perform a significance test. There are several threats to the statistical validity, such as low statistical power, which typically results from having too few participants in the study. If a statistically significant difference between the groups is detected, it is possible to move on to discuss other forms of validity. Thus, statistical validity is a prerequisite for the other types of validity. If the experimental and control groups show no significant differences, then there is little point in discussing whether the findings can be generalized to other subjects.

2.9.2 INTERNAL VALIDITY

Internal validity has to do with causality. Is it possible to draw firm conclusions about cause and effect based on the study? The best way to ensure this is to have an experimental design with a control group. Participants are distributed randomly between the two conditions; this allows us to conclude that any difference observed between the groups can be attributed to the treatment. If a control group is not used and instead a pretest/posttest design is used (participants are tested before and after the course), the internal validity is threatened because one cannot exclude other causes that may have produced or caused the change. It is possible that at the same time that they invited the employees to participate in the course, the airline also implemented other changes that contributed to the observed effect. The longer the time period between the two test periods is, the more likely it is that other things can occur and produce a change. There may be situations where a pretest/posttest design can be justified—for example, as a preliminary study of a new intervention before the intervention is implemented and examined in a larger experimental study.

Even though a true experiment is optimal, it may sometimes be possible to use other designs—so-called quasi-experimental designs—to conduct valid research.

2.9.3 CONSTRUCT VALIDITY

If both the statistical and internal validity have been addressed, it is possible to examine construct validity and external validity. Construct validity addresses whether one can generalize the relationship between cause and effect from the measured variables to the constructs. Suppose that we have implemented a CRM course for air traffic controllers, and the researcher is interested to see whether the course has increased job engagement in this group. An experimental design is used and the researcher has detected a significant difference in job engagement between the two groups after the course is completed. Both statistical and internal validity are considered to be sufficient and no major threats to internal and statistical validity have been detected. The question is then whether the findings can be generalized to the constructs—that is, to CRM courses in general and to the construct of job engagement, which would involve that the specific measurement instrument actually measured job engagement.

In addition, we must ensure that the CRM course has been conducted according to the course plan and with the specified content because we want to conclude that the specific training of the CRM course gives the effect, rather than just any course. Many interventions, such as a course, could include many effective ingredients. To investigate this further, we could also evaluate other training courses that focus only on safety in aviation and do not have a focus on collaboration and communication, such as most CRM courses do. If both types of courses are effective in relation to increasing the engagement, then the specific academic content in the CRM course is not the effective ingredient, but rather attending a course in general. Another possible outcome is that there is a difference between the two courses and that the CRM group had the highest score on engagement, and the control group and the other group (safety course) scored lower on engagement. Then it would be safe to assume that the specific CRM course is the effective treatment, rather than just any course.

Construct validity in relation to study design is similar to construct validity in testing. The difference is that test validity is usually related to how well one construct is measured. However, when we talk about construct validity in study design, it is usually two (or more) constructs that are operationalized. We want to be able to make conclusions regarding the causal relationship between them rather than the measurement of one construct.

2.9.4   External Validity

External validity is the extent to which we can generalize the results to other groups, to other situations, and over time. For example, will the effect of participating in a course last over time? Could the course be applied in a different airline and perhaps for other professions? Often, not all of these questions can be answered in a single study, for a variety of reasons:

The study may have been limited to one group.
The study may have been conducted in a single organization or country.
There are limits to how long participants may be followed after the study is completed.
The experiment may have been conducted in a laboratory setting.

An example would be determining how long it takes for a group of people to evacuate an airplane cabin under two conditions (high or low reward). If we observe differences between the groups, the question is whether this difference will also apply in a critical situation with actual passengers. Probably, some factors will be the same, but there will also be differences. It is a paradox that if one is using a design where good internal validity is ensured (i.e., a randomized controlled experiment in a laboratory setting), then the external validity may suffer (i.e., it may be more difficult to generalize the findings to a real-life situation). In other words, a design will therefore often represent a compromise between different concerns, and it is rare that a design is optimal in relation to all forms of validity.

In order to document that the findings can be generalized to different settings, to different persons, and over time, it is necessary to perform many studies where these aspects are varied. An effective way to systematize and compare several studies is through a meta-analysis.

2.10   META-ANALYSIS

Meta-analysis involves using statistical techniques to combine results from several studies that address the same problem. Suppose a researcher is interested in examining whether psychological tests measuring spatial abilities can be used to select pilots. How can this problem be investigated? One solution is to perform a so-called primary study. This means that a study that examines this problem is conducted. For example, a group seeking flight training is tested using a measure of spatial ability, and then later these results are combined with information on performance. If the research hypothesis is correct, then those with the highest test scores would receive the highest ratings by the instructor.

Another possibility would be to review other studies that have previously examined this issue. For most topics, published studies addressing the research question range from a handful to many hundreds of studies. If the number of studies is large, it can be difficult to get a clear picture of the overall results. In addition, studies probably vary in terms of samples used and measurement instruments, and it may therefore be difficult to summarize everything simply by reading through the articles.

An alternative to a narrative review of the studies would be to do a meta-analysis. In this approach, all the studies are coded and an overall measure of effect is recorded (such as ES or r) from each article or report. In this example, it will probably be a correlation between test results and instructor ratings. The meta-analysis would involve calculating a mean correlation based on all studies, and the next step would be to study variation between studies. In some studies, a strong correlation may be detected, but in others no correlation between test results and performance will be found.

A meta-analysis consists of several steps similar to the research process in a primary study:

* formulating a research problem;
* locating studies;
* coding studies;
* meta-analysis calculations; and
* presenting the results.

2.10.1   LITERATURE SEARCH AND CODING OF ARTICLES

Locating studies for a meta-analysis usually starts with a literature search using available electronic databases. This can be PsychInfo, which includes most of the articles published in psychology, or Medline, which contains the medical literature. They are many different databases depending on the discipline in which one is interested. It is thus important to use the right keywords so that all relevant studies are retrieved. In addition to searching in these databases, studies that are published in the form of technical reports or conference presentations may also be used. Often these can be found on the Web—for example, via the Web sites of relevant organizations. In addition, reference lists of articles that have already been obtained could provide further studies. Basically, as many studies as possible should be collected, but it is, of course, possible to limit the meta-analysis to recent studies or where a particular occupational group is examined.

Then the work of developing a coding form starts, and the form should include all relevant information from the primary studies. Information about the selection of participants, age, gender, effects, measurement instruments, reliability, and other study characteristics may be included. The coding phase is usually the most time-consuming part of a meta-analysis; if many studies need to be encoded, more than one coder is usually needed. Coder reliability should be estimated by having a sample of studies coded by two or more coders.

2.10.2 STATISTICAL SOURCES OF ERROR IN STUDIES AND META-ANALYSIS CALCULATIONS

Hunter and Schmidt (2003) have described a number of factors or circumstances that may affect the size of the observed correlation (or effect size). One such factor is the lack of reliability in measurements. This will cause the observed correlations to be lower than if measurements had been more reliable. These statistical error sources or artifacts will cause us to observe differences between the studies in addition to variations caused by sampling error. When a meta-analysis is conducted, the effect sizes should be corrected for statistical errors; in addition, the observed variance between studies should be corrected for sampling error. Some of these error sources will be addressed in greater detail in Chapter 4 on selection. For simplicity, we will further limit this presentation to a "bare-bones" meta-analysis where only sampling error is taken into account.

The average effect size is calculated as a sample size weighted mean. Thus, studies based on a large number of people are given more weight than those based on smaller samples. The formula for calculating the mean effect size is

$$\bar{r} = \frac{\sum N_i r_i}{\sum N_i}$$

$$\overline{ES} = \frac{\sum N_i ES_i}{\sum N_i}$$

where
  N = sample size
  r = correlation
  ES = effect size

True variance between the studies is calculated as the difference between the observed variance between effect sizes and the variance due to sampling error—that is, random errors caused by studying a sample and not the entire population.

The formula for estimating the population variance between the studies is

$$\sigma^2_\rho = \sigma^2_r - \sigma^2_e$$

where

σ^2_ρ = population variance between correlations

σ^2_r = observed variance between correlations

σ^2_e = variance due to sampling error

2.10.3 META-ANALYSIS EXAMPLE

Suppose a researcher wants to sum up studies examining the relationship between an ability test and job performance for air traffic controllers. Suppose 17 studies reporting correlations and the corresponding sample sizes are available (see Table 2.5). The data set is fictitious, but not completely unrealistic. The most important part will be to calculate the mean weighted correlation as a measure of the test's predictive validity. In addition, it will be interesting to know whether there is some variation between studies or, to put it in a slightly different way: To what extent can the predictive validity be generalized across studies?

The calculations (presented in Figure 2.3) show that the average validity is .35 (unweighted), while a sample size weighted mean is slightly lower (.30). This means that there is a negative correlation between sample size and the correlation, which

TABLE 2.5
Data Set with Correlations

Study	N	r
1	129	.22
2	55	.55
3	37	.44
4	115	.20
5	24	.47
6	34	.40
7	170	.22
8	49	.40
9	131	.20
10	88	.37
11	59	.28
12	95	.26
13	80	.42
14	115	.25
15	47	.38
16	30	.40
17	44	.50

Bare-bones meta-analysis calculations:

Number of substudies	: 17
Number of correlations	: 17
Mean number of cases in sub-studies	: 76
Mean number of cases in corr.	: 76
Total number of cases	: 1302
R Mean (Weighted)	: 0.30260
R Mean (Simple)	: 0.35059

Observed variance	: 0.01110
Std. dev of	: 0.10538

Sampling variance	: 0.01092
Std. dev of	: 0.10449

Lower endpoint 95% CRI	: 0.27580
Upper endpoint 95% CRI	: 0.32939
Credibility value 90%	: 0.28510

Population variance	: 0.00019
Std. dev of	: 0.01367

Percentage of observed variance accounted for by sampling error	: 98.32%

95% confidence interval (Homogeneous case)	
Lower endpoint	: 0.5293
Upper endpoint	: 0.35227

95% confidence interval (Heterogeneous case)	
Lower endpoint	: 0.25250
Upper endpoint	: 0.35269

Analysis performed with Metados
(Martinussen and Fjukstad, 1995)

FIGURE 2.3   Bare-bones meta-analysis calculations.

implies that studies with small sample sizes have somewhat higher correlations than those with higher n. If we correct the observed variance for sampling error, the remaining variance almost equals zero (based on the output in Figure 2.3: 0.01110 – 0.01092 = 0.00019). This means that there is no true variation between studies. Thus, the average correlation is a good measure of the predictive validity of this test.

2.10.4   CRITICISM OF THE META-ANALYSIS METHOD

Currently, the several traditions within meta-analysis employ somewhat different techniques to sum up and compare studies. The main difference concerns how the various studies are compared and whether inferential statistics are used to examine differences between studies. The alternative to this is the procedure suggested by Hunter and Schmidt (2003) in which emphasis is put on estimating variance between studies.

A problem that has been raised in relation to meta-analysis is whether the published studies can be said to be a representative sample of all studies conducted. Perhaps there is a bias in which published studies are systematically different from unpublished studies. It is reasonable to assume that studies with significant findings have a higher likelihood of being published than studies with no significant results. Because a meta-analysis is largely based on published studies, it is reasonable to assume that we overestimate the actual effect to some extent. Statistical methods can calculate the number of undiscovered studies with null effects that would have to exist in order to reduce the mean effect to a nonsignificant level. If this number is very large (e.g., 10 times the number of retrieved studies), then it is unlikely that such a large number of undiscovered studies would exist, and the overall effect can be trusted.

Another criticism of meta-analysis has been that studies that are not comparable are summarized together or that categories that are too global are used for the effects. The research problem should be considered when making coding decisions including which categories and methods should be used. For example, if we are interested in the effect of therapy in reducing PTSD (posttraumatic stress disorder), then combing different types of treatment would make sense; however, if the researcher is interested in discovering whether cognitive behavioral therapy works better than, for example, group therapy, then obviously categories for each therapy form will be needed. This problem is known in the literature as the "apple and orange" problem. Whether it is a good idea to combine apples and oranges depends on the purpose. If one wants to make fruit salad, it can be a very good idea; on the other hand, if one only likes apples, the oranges are best avoided.

2.11   RESEARCH ETHICS

Research has to comply with many rules and regulations in addition to scientific standards—for example, research ethics. Research ethics include how participants are treated and the relationship between the researcher and other researchers, as well as relationships with the public. International conventions, as well as national regulations and laws, govern research ethics. Each country and sometimes even large organizations have their own ethics committees where all projects are evaluated. One such international agreement is the Helsinki Declaration, which includes biomedical research conducted on humans. According to the declaration, research should be conducted in accordance with recognized scientific principles by a person with research competence (e.g., a PhD), and the subjects' welfare, integrity, and right to privacy must be safeguarded. Informed consent to participate in the study should be obtained from the subjects before the study begins.

In general, there is substantial agreement on the basic principles that should govern research, although the formal approval procedures to which projects are subjected may vary slightly from country to country. People who participate in research projects should be exposed to as little discomfort or pain as possible, and this must be carefully weighed against the potential benefits of the research. These two perspectives—society's need for knowledge and the welfare of the participants—need to be balanced. Participation should be voluntary, and special care needs to be taken when people are in a vulnerable position or in a special position in relation to the researcher (e.g., a subordinate or a client).

An important principle in relation to research ethics is *informed consent*. This means that the person asked to participate should have information about the purpose of the study, the methods to be used, and what it will involve in terms of time, discomfort, and other factors to participate in the study. The person should also receive information about opportunities for feedback and who can be contacted if additional information is needed. It should also be emphasized that participation is voluntary and that information is treated confidentially. For many studies involving an experiment or an interview, it is common for people to sign a consent form before the study begins. If the study is an anonymous survey, then a consent form is not usually attached; instead, the person gives consent to participate implicitly by submitting the questionnaire. The person should also receive information that he or she may at any time withdraw from the study and have his or her data deleted.

In some studies, perhaps especially from social psychology, subjects have been deceived about the real purpose of the study. An example is experiments where one or more of the research assistants act as participants in the experiment; the purpose is to study how the test subjects are influenced by what other people say or do. The most famous experiment in which subjects were deliberately misled about the true purpose of the experiment was the Milgram studies on obedience conducted in the 1960s. This study was presented as an experiment in learning and memory, but it was really an experiment to study obedience. Subjects were asked to punish with electrical shocks a person who in reality was a research assistant. Many continued to give electric shock even after the person screamed for help. Many of the subjects reacted with different stress responses and showed obvious discomfort in the situation, but nevertheless continued to give shocks.

These studies violate several of the ethical principles outlined in this section. These include lack of informed consent, the fact that people were pressured to continue even after they indicated that they no longer wanted to participate, and exposing people to significant discomfort even though they were informed about the purpose of the study afterward.

Such experiments would probably not be approved today, and a researcher would need to make a strong argument for why it would be necessary to deceive people on purpose. If the researcher did not inform the participants about the whole purpose of the project or withheld some information, the participants would need to be debriefed afterward. A common procedure in pharmaceutical testing is to provide a group of people with the drug while the other group (control) receives a placebo (i.e., tablets without active substances). Such clinical trials would be difficult to perform if the subjects were informed in advance about their group assignment. Instead, it is common to inform the subjects that they will be in the experimental group or in the placebo group and that they will not be informed about the group to which they belonged until the experiment ended.

2.12 CHEATING AND FRAUD IN RESEARCH

Research dishonesty can take many forms. One of the most serious forms of fraud is tampering with or direct fabrication of data. Several examples have been published in the media, both from psychology and other disciplines where scientists have constructed all or parts of their data. Another form of dishonesty is to withhold parts of a data set and selectively present the data that fit with the hypothesis. Other types of dishonest behavior include stealing other researchers' ideas or text without quoting or acknowledging the source, or presenting misleading representations of others' results.

It should be possible to verify a researcher's results, so the raw data should be stored for at least 10 years after the article has been published, and the data should be made available to others in the event of any doubt about the findings. Many countries have permanent committees that will investigate fraud and academic dishonesty whenever needed.

2.13   SUMMARY

During the Crimean War, far more British soldiers died in field hospitals than on the battlefield due to various infections and poor hygienic conditions. Florence Nightingale discovered these problems and implemented several reforms to improve health conditions in the field hospitals. To convince the health authorities about the benefits of hygiene interventions, she used statistics, including pie charts, to demonstrate how the mortality rates of the hospitals changed under different conditions (Cohen 1984). She also demonstrated how social phenomena could be objectively measured and analyzed and that statistics were important tools to make a convincing argument for hospital reforms.

The main topic of this chapter has been related to how we can gain new knowledge and about important research requirements concerning methods, design, analyses, and conclusions. Research methods and statistics are important tools when a phenomenon is investigated and the results presented to others. Without methodology and statistics, it would be very difficult to present a convincing argument for the statements one wished to make, and the example set by the Florence Nightingale is still valid today.

RECOMMENDED READING

Hunt, M. 1997. *How science takes stock. The story of meta-analysis.* New York: Russel Sage Foundation.

Lipsey, M. W., and Wilson, D. B. 2001. *Practical meta-analysis.* Applied Social Research Methods Series, vol. 49. London: Sage Publications.

Murphy, K. R., and Davidshofer, C. O. 2005. *Psychological testing. Principles and applications,* 6th ed. Upper Saddle River, NJ: Pearson Education.

Wiggins, M. W., and Stevens, C. 1999. *Aviation social science: Research methods in practice.* Aldershot, England: Ashgate Publishing Ltd.

REFERENCES

American Psychological Association. 1999. *The standards for educational and psychological testing.* Washington, D.C.: American Psychological Association.

Borenstein, M., Rothstein, H., and Cohen, J. 2001. *Power and precision: A computer program for statistical power analysis and conficence intervals.* Englewood Cliffs, NJ: Biostat Inc.

Cohen, I. B. 1984. Florence Nightingale. *Scientific American* 250:128–137.

Embretson, S. E., and Reise, S. 2000. *Item response theory for psychologists.* Mahwah, NJ: Lawrence Erlbaum Associates.

Fink, A., and Kosecoff, J. 1985. *How to conduct surveys. A step by step guide.* Newbury Park, CA: Sage Publications, Inc.

Guion, R. M. 1981. On Trinitarian doctrines of validity. *Professional Psychology* 11:385–398.

Hunter, J. E., and Schmidt, F. L. 2003. *Methods of meta-analysis: Correcting error and bias in research findings.* Newbury Park, CA: Sage.

Magnusson, D. 2003. *Testteori* [*Test theory*], 2nd ed. Stockholm: Psykologiförlaget AB.

Martinussen, M., and Fjukstad, B. 1995. *Metados: A computer program for meta-analysis calculations.* Universitetet i Tromsø.

Messick, S. 1995. Validity of psychological assessment. *American Psychologist* 50:741–749.

Parsons, H. M. 1992. Hawthorne, an early OBM experiment. *Journal of Organizational Behavior Management* 12:27–43.

Shadish, W. R., Cook, T. D., and Campbell, D. T. 2002. *Experimental and quasi-experimental designs.* Boston, MA: Houghton Mifflin Company.

Van Horn, P., Green, K., and Martinussen, M. 2009. Survey response rates and survey administration in clinical and counseling psychology: A meta-analysis. *Educational and Psychological Measurement.* 69:389–403.

3 Aviation Psychology, Human Factors, and the Design of Aviation Systems

This, indeed, is the historical imperative of human factors—understanding why people do what they do so we can tweak, change the world in which they work and shape their assessments and actions accordingly.

Dekker (2003, p. 3)

3.1 INTRODUCTION

As noted in the introduction in Chapter 1, aviation psychology is closely related to the field known as human factors. In recent years the distinction among aviation psychology, human factors, and the more hardware-oriented discipline of engineering psychology has become very blurred, with practitioners claiming allegiance to the disciplines performing very similar research and applying their knowledge in very similar ways. Traditionally, engineering psychology might be thought of as focusing more on humans and human factors might focus somewhat more on hardware and its interface with the human operator. For all practical purposes, however, the distinction between the two disciplines is irrelevant. It is mentioned here only to alert the reader to the terminology because much of what we would label as aviation psychology is published in books and journals labeled as human factors.

Setting aside the differences in terminology, aviation psychology (or human factors) has a great deal to say about how aviation systems should be designed. To meet the goals of reducing errors, improving performance, and enhancing comfort, a system must accommodate the physical, sensory, cognitive, and psychological characteristics of the operator. A system must not demand that the operator lift excessive weights or press a control with an impossible amount of force. A system must not require that the operator read information written in a tiny font or make fine distinctions of sound when operating in a noisy environment. A system must not demand complex mental arithmetic or the memorization and perfect recall of long lists of control settings, dial readings, and procedures. A system must not demand that the operator remain immune to the social stresses placed on him or her by co-workers or to the demands of management to cut corners to accomplish the job.

Knowledge of human capabilities, strengths, and limitations informs the system design process because this knowledge sets the bounds for the demands the system may make of the operator. An extensive body of research has addressed these bounds. Researchers have studied how much weight humans can lift to specified heights, the numbers of errors that occur when identical controls are placed side by side, how many numbers can be recalled from short-term memory, the font size of displays, legibility of displays under varying degrees of illumination, and the effects of an organization's climate on the safety-related behavior of workers, to list but a few examples. The overall aim of this chapter is to demonstrate how psychological knowledge may be used when designing aviation systems, what principles should be applied, and common errors and problems that occur when humans interact with complex systems and equipment.

3.2   TYPES OF HUMAN ERROR

Arguably, the present status of aviation psychology and human factors owes much to the efforts of researchers during World War II. The sheer magnitude of the war effort led researchers on both sides of the conflict to conduct extensive studies with the aim of improving personnel performance and reducing losses due to accidents and combat. Perhaps the most frequently cited study in the area of aviation psychology and human factors produced by that era was the work by Fitts and Jones (1947a) on the causes of errors among pilots.

Fitts and Jones (1947a, 1961a) surveyed a large number of U.S. Army Air Force pilots regarding instances in which they committed or observed an error in the operation of a cockpit control (flight control, engine control, toggle switch, selector switch, etc.). They found that all errors could be classified into one of six categories:

- substitution errors—confusing one control with another or failing to identify a control when it was needed;
- adjustment errors—operating a control too slowly or too rapidly, moving a switch to the wrong position, or following the wrong sequence when operating several controls;
- forgetting errors—failing to check, unlock, or use a control at the proper time;
- reversal errors—moving a control in the direction opposite to that necessary to achieve the desired result;
- unintentional activation—operating a control inadvertently without being aware of it; and
- unable to reach a control—inability physically to reach a needed control or being required to divert attention from an external scan to such a point that an accident or near-accident occurred.

Substitution errors accounted for 50% of all the error descriptions reported; the most common types of errors were confusion of throttle quadrant controls (19%), confusion of flap and wheel controls (16%), and selection of the wrong engine control or propeller feathering button (8%). The conditions that gave rise to such results are illustrated in Table 3.1, using data provided by Fitts and Jones (1961a, p. 339) on the throttle quadrant configurations on three common aircraft of that era. Similar

TABLE 3.1
Aircraft Configurations Leading to Control Confusion

Aircraft	Control Sequence on Throttle Quadrant		
	Left	Center	Right
B-25	Throttle	Propeller	Mixture
C-47	Propeller	Throttle	Mixture
C-82	Mixture	Throttle	Propeller

difficulties were encountered with the controls for the flaps and landing gear, which at that time were often located close to one another and used the same knob shape.

Fortunately for today's pilots, many of the recommendations of Fitts and Jones and other researchers of that period have been implemented. The configuration of the six principal instruments, the order of controls on the throttle quadrant for propeller-driven aircraft, and the shapes of the controls themselves are all now fairly standardized. The shape of the knob for the landing gear resembles a wheel, the shape of the flaps knob resembles an airfoil, and the two controls are located as far apart as possible while still remaining easily accessible to the pilot.

Although these sorts of errors have been largely, though not entirely, eliminated, others remain. "Forgetting" errors, which in the Fitts and Jones study accounted for 18% of the total errors, remain a problem in today's aircraft. The shape of the landing gear control may have largely prevented its confusion with the flaps; however, the pilot must still remember to lower the gear prior to landing. Memory devices, paper checklists, and, in the case of more advanced aircraft, computer watchdogs all serve to prevent the pilot from making the all-too-human error of forgetting. Interestingly, one of the recommendations of Fitts and Jones (1961a, p. 333) was to make it "impossible to start the takeoff run until all vital steps are completed." Clearly, this is a goal that still eludes us: Pilots still attempt takeoffs without first extending the leading-edge slats and flaps, and they make landings without prior arming of the spoilers—typically, after defeating the warning systems put in place to prevent such events.

3.3 HUMAN CHARACTERISTICS AND DESIGN

At a more general level than the work by Fitts and Jones, Sinaiko and Buckley (1957; 1961, p. 4) list the following general characteristics of humans as a system component:

- physical dimensions;
- capability for data sensing;
- capability for data processing;
- capability for motor activity;
- capability for learning;
- physical and psychological needs;
- sensitivities to physical environment;

- sensitivities to social environment;
- coordinated action; and
- differences among individuals.

All of these characteristics must be taken into account in the design of aviation systems. Some of the system requirements driven by these characteristics are reasonably well understood and have been addressed in system design for many years. For example, certainly since the work of Fitts and Jones following World War II, designers have been aware of the need to mark and separate controls properly and to arrange displays in a consistent way. However, the implications of some of the characteristics are still being explored. The work over the past 20 years on crew resource management (see Helmreich, Merritt, and Wilhelm, 1999, for an overview) is evidence of our growing understanding of the sensitivities of humans to their social environment and capabilities for coordinated action. Even more recently, researchers have begun to explore the influences of the organizational climate and culture on the performance of aircrew (Ciavarelli et al. 2001).

Of particular relevance to aviation psychology is the notion of differences among individuals. Although Sinaiko and Buckley (1957) list it as a separate characteristic, it is really inherent in all the other characteristics they list. Humans vary, often considerably, on every characteristic by which they may be measured. The measurement of these individual differences and determination of how they contribute to other characteristics of interest—such as success in training, likelihood of an accident, skill at making instrument landings, or probability of being a good team member—are at the heart of aviation psychology.

In addition to examining the errors associated with controls, Fitts and Jones (1947b, 1961b) also examined errors in reading and interpreting aircraft instruments. As in their study of control errors, they classified errors in reading or interpreting instruments into nine major categories. Errors in reading multirevolution instrument indications accounted for the largest proportion of errors (18%). Misreading the altimeter by 1,000 feet was the most common of these errors, accounting for 13% of the total errors. Additional errors included reversal errors (17%), signal interpretation errors (14%), legibility errors (14%), substitution errors (13%), and using inoperative instruments (9%).

Among their several conclusions, Fitts and Jones (1961b, p. 360) noted that "the nature of instrument-reading errors is such that it should be possible to eliminate most of the errors by proper design of instruments." Arguably, 60 years after the first publication of their work, current researchers could still arrive at the same conclusion. If most of the issues associated with the shape of controls have been resolved, problems with displays remain. However, they are not necessarily the same problems identified by Fitts and Jones. Multirevolution instruments (most notably the altimeter) have been replaced with instruments that depict information differently—typically, along a vertical scale in the case of altitude. Yet, pilots still fly into the ground on occasion because, even after reading the instrument correctly, they have misprogrammed the system that controls the vertical flight profile of their aircraft. Likewise, radio navigation beacons required pilots to identify the Morse signal transmitted by the beacon aurally and gave rise to signal interpretation errors; these have given way to GPS navigation, with its own set of display problems and corresponding errors.

Each new generation of technology offers some solutions to the problems that existed in the older generation, while creating a whole new set of problems. This situation is succinctly described by Dekker (2002, p. 8), who notes that "aerospace has seen the introduction of more technology as illusory antidote to the plague of human error. Instead of reducing human error, technology changed it, aggravated the consequences and delayed opportunities for error detection and recovery."

3.4   PRINCIPLES OF DISPLAY DESIGN

One way to break this chain of technology and error is to step outside the specific technologies and look at overarching principles that should be applied to all new technology development. Thus, instead of looking for the best shape for the landing gear control, we might look for the general principles by which such controls should be designed. As an example, let us consider the design of aircraft displays. Wickens (2003), one of the preeminent researchers in the area of aviation displays, has enumerated seven critical principles of display design, which are described next.

3.4.1   PRINCIPLE OF INFORMATION NEED

How much information does a pilot need? The short answer is "just enough." Too little information (e.g., the absence of weather radar on days when thunderstorms are present) leaves the pilot flying, and making decisions, blind. Most pilots would agree that having more information is good, but the converse is also true. Having too much information can be as damaging as having too little. Too much information can lead to a cluttered flight deck (typified by the L-1011 and DC-10 era aircraft) with hundreds of dials and indicators. Searching for the needed information among all the extraneous information can lead to poor performance on critical, time-sensitive tasks. Current-generation aircraft, in contrast, have combined many of the formerly separate information sources into combined displays that integrate information, such as engine health, into a single, easily interpretable instrument. When that information is needed, it is easily obtained.

To determine how much information is enough, we turn to a family of techniques subsumed under the title "task analysis"* (cf. Kirwan and Ainsworth 1992; Meister 1985; Seamster, Redding, and Kaempf 1997; Shepherd 2001; Annett and Stanton 2000). Although several varieties of this technique exist and sometimes are used for different purposes (e.g., for training development or for personnel selection, as mentioned in other chapters of this book), they share a general approach to the orderly specification of the tasks that a person must accomplish, the actions (both physical and cognitive) that the person must complete, and the information required to permit

---

* This brief discussion cannot hope to do justice to a topic that is the subject of many volumes. The reader is encouraged to consult the general references to task analysis listed here for more information. However, even these are only a tiny sampling of the vast amount of information available. Because task analysis methods vary according to the intended use of the information, the survey of methods given in Annett and Stanton (2000) may prove most beneficial for the task analysis novice.

the person to complete the actions. For example, we might specify the information required to complete a precision instrument approach or the information required to identify which of several engines has failed. If the pilot is expected to complete these tasks (making the instrument approach and dealing with the failed engine), then he or she must have the required information. In addition, that information should not be hidden by or among other bits of information.

3.4.2   PRINCIPLE OF LEGIBILITY

In order to be useful, information presented on displays must be readable. Further, it must be readable under the conditions that exist in the aircraft flight deck. This means that the digits on the display must be large enough to be read by the pilot from his or her normal seated position. In some cases, they should also be readable by the other crew member—for example, if there is only one such display on the flight deck and both crew members must use it. Typically, designers solve this problem by locating common displays and controls midway between the two pilots on a central console.

Legibility also requires consideration of effects such as glare and vibration. For example, almost all pilots quickly learn to steady their hands when reaching to tune the radio because even mild turbulence can make such a task quite difficult to perform rapidly and accurately. This vibration also impacts the legibility of displays, and the usual solution is to incorporate a larger font size so that the information can be read under the full range of operating conditions. The effects of factors such as these on human performance have been extensively investigated and are summarized by Boff, Kaufman, and Thomas (1988). In addition, the Engineering Data Compendium is an extensive online resource from the Human Systems Integration Information Analysis Center of the U.S. Air Force.* Readers may also find Sanders and McCormick (1993) and Wickens and Hollands (2000) useful references on this topic.

3.4.3   PRINCIPLE OF DISPLAY INTEGRATION/PROXIMITY COMPATIBILITY PRINCIPLE

The novice pilot, particularly the novice instrument pilot, has no doubt that scanning the instruments to obtain the information required to control the aircraft and navigate requires effort. The level of effort required can be increased or decreased by the degree to which the instruments are physically separated. Thus, in all modern aircraft the primary flight instruments are located directly in front of the pilot. This reduces the time required for the pilot to move his or her scan from one instrument to another. It also means that the instruments can be scanned without moving the head—thus reducing the potential for vestibular disorientation.

In addition, effort can be reduced if displays that contain information that must be integrated or compared are close together. This is seen most clearly in multiengine aircraft where two, three, or four sets of engine instruments are arrayed (in older aircraft) in columns, with each column corresponding to one engine and each row to one engine parameter (e.g., oil temperature or turbine speed). Given this arrangement,

---

* http://www.hsiiac.org/products/compendium.html

the pilot may quickly scan across all the engine temperature readings, for example, to identify an engine with an anomalous reading.

Further examples are evident in the navigation instruments. For example, the lights indicating passage of marker beacons during a precision instrument approach are typically located close to the primary flight control instruments and the instrument landing system (ILS) display. Having the lights in the direct field of view of the pilot, instead of somewhere in the radio stack, enhances the likelihood that they will be seen by the pilot. This is particularly important for these displays because they usually are extinguished after passage, with no persistent indicator that an important event has transpired.

Integration of related information into a single instrument represents a means to further reduce pilot workload by eliminating the necessity to scan multiple instruments visually and, potentially, eliminating the necessity to combine separate bits of information cognitively. Perhaps the best example of this integration is the display for the flight management system (FMS) on a modern transport category aircraft. This system brings together in one display (typically called the primary flight display) virtually all the information required for control of the aircraft and for horizontal and vertical navigation.

At a somewhat simpler level, the flight director built into the attitude displays of some general aviation aircraft illustrates the same principles of integration. The flight director provides visual cues on the attitude indicator. In the simplest form, these cues can take the form of simple horizontal and vertical lines—depicting the localizer and glideslope for an ILS, for example. Essentially, this arrangement moves these indicators from the ILS display to the attitude indicator, thus eliminating the need for the pilot to move his or her scan between these two instruments. Another configuration makes use of a black inverted "V" that represents the visual cue from the flight director, as shown in Figure 3.1.

In this figure, the triangle represents the aircraft on the attitude indicator. In this situation, the flight director cues indicate that the pilot needs to bank the aircraft to the left. If the pilot keeps the triangle tucked up into the black "V," he or she will satisfy the cues from the flight director and will follow the desired course (i.e., ILS localizer/glideslope, VOR [very high frequency omnidirectional radio] radial, or GPS [global positioning system] course).

3.4.4 Principle of Pictorial Realism

This principle holds that the display should resemble or be a very similar pictorial representation of the information it represents. The moving tape that represents altitude by moving a tape vertically represents one application of this principle. The current generation of moving map displays that can show the location of the aircraft (often depicted with a small aircraft symbol) against a background of topographic imagery is arguably an even stronger example of this principle.

3.4.5 Principle of the Moving Part

According to this principle, the element that moves on a display should correspond to the element that moves in a pilot's mental model of the aircraft. In

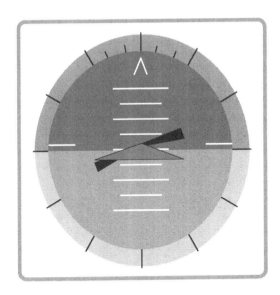

FIGURE 3.1   Typical general aviation flight director display.

addition, the direction of movement of the display element should correspond with the direction of movement in the mental representation. Arguably, this principle is best illustrated by the instrument that most thoroughly violates the principle: the attitude indicator. In the attitude indicator, the horizon is depicted as a moving element, whereas the aircraft is shown as static. However, this is completely opposite to the pilot's mental model, in which the horizon is static and the aircraft banks, climbs, and descends. The sacrifice in human performance demanded by this arrangement is reflected in the finding that, for novice pilots, the moving aircraft display is more effective than the moving horizon display. Furthermore, even for pilots who are experienced in flying with the traditional, moving horizon display, the moving aircraft display is no less effective (Cohen et al. 2001; Previc and Ercoline 1999).

3.4.6   Principle of Predictive Aiding

Predicting the future state of the aircraft (heading, altitude, rate of climb or descent, bearing to some beacon, etc.) is a complex and cognitively demanding task. Insofar as possible, displays should assist the pilot in this task by showing what will happen in the future. This allows the pilot to take steps now so that the desired state is achieved or an undesirable state is avoided. Many of the current generation of FMSs provide this service by showing predicted flight paths, based on current engine and control settings.

However, valuable assistance may be obtained from far less sophisticated systems. Consider the example of the fuel gauge, which we will discuss in more detail later in this chapter. Most current designs simply represent the current status of the fuel supply—somewhere between full and empty. However, a slightly more sophisticated gauge could show future states, such as when and/or where zero fuel remaining will be reached, based on the current fuel load and consumption rate. This simple predictive aiding might help prevent the 10% of all accidents due to fuel mismanagement.

3.4.7 PRINCIPLE OF DISCRIMINABILITY: STATUS VERSUS COMMAND

Preventing confusion among similar displays is a responsibility of designers. Unfortunately, engineering demands often lead to sacrifices in this area. One example is the use of identically sized displays for all the engine instruments because commonality reduces cost. This arrangement saves money during manufacturing because only one size of hole need be punched in the panel; however, this can later lead to a pilot mistaking one instrument for another, with results that can range from humorous to disastrous.

Unambiguous information is essential for the safe operation of the aircraft. Particularly problematic are those instances in which similar information, with an entirely different meaning, is displayed in a common display. This is a condition that is not unknown in FMSs and has been cited as the cause of at least one major crash (Air France Airbus A-320 that crashed in Mulhouse-Habsheim Airport, France).

In addition to Wickens, many other researchers have also evaluated display issues. For example, looking specifically at the symbols used in displays, Yeh and Chandra (2004) posed four questions to be addressed when evaluating the usability of a symbol:

- Is the symbol easy to find?
- Is the symbol distinctive from other symbols?
- Is the on-screen symbol size appropriate?
- Can all encoded attributes of the symbol be decoded quickly and accurately?

Much of the work up through the mid-1980s is summarized in Boff et al. (1988) and is readily available through the Engineering Data Compendium,* an online source of data related to human performance and design issues.

The principles espoused by Wickens and others in the design of aviation systems, along with the results of many empirical studies on the effects of system characteristics on human performance, are codified in government regulations pertaining to the design of aircraft control and display systems. Particularly detailed listings of design standards are also provided in the military standards and handbooks used to govern the design and development of new military aircraft and related systems. Indeed, much of the development of the knowledge relating to human capabilities and the corresponding standards for system design has been led by the military. One recent example of the military's efforts to improve the design process is the U.S. Army's MANPRINT program (Booher 1990, 2003).

---

* http://www.hsiiac.org/products/compendium.html

3.5 SYSTEM DESIGN

MANPRINT stands for "manpower, personnel, and training integration"; however, MANPRINT also subsumes human factors, systems safety, and health hazards concerns. This program, which is codified into an army regulation, expresses the U.S. Army's philosophy of a soldier-centric design process, in which the design of new military systems (e.g., new helicopters) starts from a consideration of who will operate (pilot) and maintain (aircraft mechanics) the new system.

3.5.1 MANPOWER

In the MANPRINT program, manpower expresses the number of people who will be required to operate and maintain the system. Clearly, this is a major concern for the military because a system that requires additional people, all things being equal, will cost more to operate than a system that requires fewer people. The same concern arises in nonmilitary settings. For example, the design of two-person flight decks in the current generation of transport aircraft represents a considerable savings to airlines over the previous flight decks that required three (or more) operators. Consider, however, the impact of such a design decision.

Changing from a three-person to a two-person crew requires much more than simply moving all the deleted person's controls and displays to one of the other crew members. It requires, first, a careful consideration of the tasks that the third person performed, the tasks that the other persons performed, and the potential for having the system perform some or all of these tasks. If these tasks were essential for the safe and efficient operation of the aircraft in a three-person crew configuration, then they will still need to be performed in the two-person configuration. However, now they must be performed by one of the remaining crew members or by some level of automation. For the most part, the current generation of aircraft has relied upon the third option, automation, to compensate for the missing crew member. This approach is not without its own peculiar drawbacks, however, because adding automation adds new tasks that are often quite different, and possibly more difficult, than the tasks that were eliminated. This problem will be illustrated in more detail later. For now, let us state one general rule: *All designs begin from an understanding of the tasks to be performed.* With that summary, let us return to our discussion of MANPRINT.

> Task analysis: A task analysis is a process of documenting the steps a user is required to perform (actions and/or cognitive processes) to complete a task.

3.5.2 PERSONNEL

Knowing the tasks to be performed allows the second element of MANPRINT to be assessed. "Personnel" refers to the qualifications and characteristics of the people who will operate and maintain the system. At its most rudimentary level, it refers to the general mental capacity or intelligence of the operators and maintainers. Usually, complex tasks require more mental capacity than simple tasks. However, the tasks required to be performed by the operator and maintainer may also demand physical strength (e.g., the mechanic may need to lift heavy objects above his or her head while making certain repairs), color vision (air traffic controllers and pilots must be able to distinguish the colors of the color gun used by the tower during radio failures), or spatial reasoning ability (required of pilots and air traffic controllers to understand the potential for traffic conflicts), to give a few examples.

Taken together, these personnel characteristics required for task performance amount to a specification for the human operators and maintainers, in much the same sense as the specifications given for the materials used to construct the airframe or power plant. In the same sense, using substandard materials that do not meet the specifications can lead to poor performance or failure. Ensuring that the human components meet these specifications involves personnel selection and training, both of which are addressed in other chapters of this book.

3.5.3 Training

Training, as indicated earlier, is the third major component of the MANPRINT program. Regardless of one's attitudes about the military or military service, it is undeniable that the military services are consummate practitioners of the art and science of training. Although the numbers of trainees and the quality of the training certainly vary from nation to nation, among the Western democracies with which the authors are familiar, the training programs are most impressive. Consider, for example, the challenge of taking a person who has never flown an aircraft (perhaps has never even been in an aircraft) and in the space of about a year, transforming that person into a fully qualified military pilot, capable of operating a sophisticated aircraft under conditions considerably more demanding that those faced by civilian counterparts. This is a most impressive accomplishment, and it is made possible by a highly structured approach to the incremental accumulation of knowledge by the aspiring pilot. (See Chapter 5 on training for a detailed description of the systems approach to training used by the services.)

Of course, this training program must be appropriate for the system that the trainee will eventually operate. A training program that only taught, for example, navigation by the use of nondirectional radio beacons or even VOR, without mention of GPSs or distance measuring equipment (DME), would likely produce graduates who were unsuited for the tasks that they must perform in an operational setting. Another general rule that we might state at this point is that *training content is driven by the equipment that will be used and the tasks that will be performed by the graduates.*

Once again we see that tasks are a central concern. In addition, however, training content and duration are also driven by the personnel entering training. The military services select high-quality personnel from off the street (these are often referred to as "ab initio," meaning "from the beginning") with no prior experience to train as pilots and mechanics. This choice of personnel means that the training schools must teach the new trainees absolutely everything they need to know to operate or maintain an aircraft. Nothing can be assumed to be known when they report for training. Contrast this with the process used by most commercial airlines, which select their pilots from a pool of individuals who already possess a pilot license and who may have many thousands of hours of experience. Certainly, these trainees can be assumed to know the basics or common piloting tasks, and the airlines' training programs are designed to build upon that foundation, providing aircraft-specific training leading to a type rating or training on company policies and procedures.

The interaction of manpower, personnel, and training considerations should now be clear. Each of these factors may be traded against the others in the design of a new system. For example, reducing crew size may require finding more capable crew members who can handle the additional tasks. Using lower quality personnel may mean that additional training is required to achieve the same standard of performance. Skimping on the human factors aspects of the crew interface may mean that additional crew members are required to perform the complex tasks that a better design would have eliminated.

Design, guard, warn, train. This is the chain of activities, given in order of precedence, for building human considerations into the system development process. Early changes to the design of a system are more effective and less costly than later attempts to guard against operator error, to warn operators of hazards, or to train operators to use the system despite its inherent problems. Despite its popularity, training should be viewed as the solution of last resort.

3.6   AN EXAMPLE: DESIGN OF THE FUEL GAUGE

To give a general idea of how information gained from aviation psychology influences system design, our discussion thus far has been fairly abstract. Now, we will take a look at a concrete example, the ubiquitous fuel gauge. The fuel gauge, at least in light aircraft, is possibly the simplest gauge in the cockpit, so just what can be said about it from the standpoint of aviation psychology? As it turns out, there is quite a lot.

Why worry about the fuel gauge? There are a lot of fuel-related accidents and incidents. In the United States, approximately 10% of accidents involving general aviation aircraft are attributed to fuel management, including fuel starvation (fuel not being fed to the engine) or fuel exhaustion (no fuel left) (Aircraft Owners and Pilots Association 2006). Also, about 20% of general aviation pilots report that they have been so low on fuel that they were worried about making it to an airport (Hunter 1995). One reason for these statistics may be the lowly fuel gauge—unseen, unheeded, misunderstood.

Where should it be placed? There is a limited amount of space on the panel of an aircraft, so how does one decide what to put where? Previous studies have resulted in the classic "T" arrangement of the primary flight instruments. Beyond that arrangement, which is specified in the aircraft certification requirements by most civil aviation authorities, where should all the other dials, knobs, lights, and indicators be placed?

How big should the gauge be? Because panel availability usually dictates the maximum size of the instrument, this comes down to the question of how large the indicator pointer and text should be. How should it be labeled? Many fuel gauges in light aircraft look much like their automotive counterparts, with markings for FULL, ¾, ½, ¼, and EMPTY. At first glance, these labels seem entirely satisfactory, but consider the mental effort required by the pilot to extract usable information from such a display. Let us pose the question, "Does the pilot care that the tank is half full?" We propose that the answer is no. In fact, what the pilot cares about is how much longer the aircraft can remain airborne without the engine quitting (that is, remaining hours and minutes of available fuel) or how much farther it can fly without becoming a glider (how many miles the available fuel will take us). The answer to either of these two questions will tell the pilot if he or she can safely continue the flight or whether to consider a diversion.

Certainly the answers can be derived from the current markings, but to do so the pilot must perform some intervening mental effort. First, he or she must convert the instrument reading (let us say "½") into gallons. Thus, the pilot looks at the gauge, sees that it reads ½, and, remembering the total fuel capacity from the pilots operating handbook (POH), calculates that there are now 10 gallons of fuel remaining (half of the 20 gallons of total usable fuel). Next, the pilot must compute the flight time that 10 gallons will afford. He or she must recall the fuel consumption rate of the engine at this particular pressure altitude and power setting (or find the POH and look up the data) and then compute the hours of remaining fuel by doing a little mental division. All of this must be done, of course, without substantial error and while simultaneously flying the aircraft, navigating, and communicating.

What is the alternative? From the foregoing discussion, certain aspects of an alternative design for the fuel gauge should be evident. First, the fuel gauge needs to be located at a place on the instrument panel in which it will be noticed by the pilot. Preferably, this should be very close to the normal instrument scan of the primary "T" instruments. Second, an alerting mechanism should be incorporated to draw the pilot's attention to certain prespecified conditions (e.g., reaching a specific level of remaining fuel). Third, the fonts and symbols used on the display should be of sufficient size and illumination so as to be clearly visible under all operating conditions, without interpretation error. Finally, the scaling of the gauge should be changed so that more relevant information that does not require extra mental effort to process is presented.

If a human-centered design is accomplished, then even a simple instrument such as a fuel gauge can be substantially improved so as to reduce operator error and improve performance. The key is to consider the display or control from the point of view of the operator and the underlying utility or purpose that the display or control serves. That is, what is the real need that is met by the control or display? In the case of the fuel gauge, the real purpose is to inform the pilot of how much longer powered flight can be maintained. It is not to tell the pilot how many gallons of fuel remain unconsumed. That information, while easy for the aircraft designer and manufacturer to implement, represents but the first step in the process that provides the pilot with the information that is really needed. It is incumbent upon the pilot to insist that designers and manufacturers not take the easy path, but rather create systems and hardware optimal for the operator to use—not for the manufacturer to build.

3.7 INTERACTING WITH THE SYSTEM

The importance of crafting the interaction between the system and the human operator goes beyond the fairly simple issues of display markings. It also extends to the nature, sequencing, and amount of information provided to the pilot. Short-term memory is the term applied to human memory for information presented and retained for a fairly short time span—typically, on the order of a few seconds to a very few minutes. A common illustration from aviation would be the recall and read-back of a new frequency assigned by air traffic control. Typically, the following is the sequence of events:

- ATC sends a voice radio message: "Aircraft 123, contact Center on 137.25."
- The pilot of the aircraft responds by saying, "Contact Center on 137.25, Aircraft 123."
- The pilot of the aircraft must then remember (hold in short-term memory) the values "137.25" while he or she reaches down and turns the radio frequency selector knobs to the appropriate setting.

Between the time that ATC says "137.25" and the time the pilot completes the action of switching the radio, a period of approximately 5–10 seconds may elapse. During that time, the pilot must keep the value "137.25" in his or her short-term memory. Usually, this is done without error, although on occasion pilots will make mistakes and enter the wrong frequency. One reason this happens only rarely is that both the span of information to be recalled and the length of time are short relative to the capacity of humans. In this example, the span is five digits.

Previous research has demonstrated that the short-term memory capacity of humans is around seven digits. The best known study of this phenomenon (Miller 1956) refers to this as the "magic number 7, plus or minus 2." As the number of digits to be recalled exceeds this "magic number," the rate of errors increases rapidly. For this reason, well-designed systems avoid requiring humans to hold more than seven digits (or other bits of information, such as words) in their short-term memory. Telephone numbers, for example, seldom exceed seven digits and, in addition, take advantage of chunking to improve recall. "Chunking" refers to grouping of the digits so as to make them more memorable. For example, instead of listing a number as 1234567, the number is given as 123-4567, or 1 23 45 67. Both of the latter arrangements are much less susceptible to recall errors.

Two other psychological phenomena that influence the recall of information are serial position effect and confirmation bias. First, let us consider serial position effect. When humans learn a list of words or other information, it has been found that they tend to recall the first and last items best. That is, they will recall the first word in the list and the last word in the list better than those that appeared in the middle. Consider how this might be important to a pilot when receiving a weather briefing.

The research has shown that he is much more likely to recall the first and last things present than those in the middle. Might the weather service take this into account by placing the most important information (perhaps the information most critical to the safety of the flight) at the start and end of the briefing? This would maximize the likelihood of the pilot's recalling this critical information.

But, it is not just the position of information that affects the pilot's recall. The predisposition of the pilot to receive information also comes into play. Psychological research has shown that humans tend to look for information that confirms or supports their pre-existing beliefs or views of the world. This tendency is called confirmation bias. In the context of the weather briefing example just given, consider how this confirmation bias might influence the pilot in recall or acceptance of information. If the pilot has already decided that the weather is adequate for the flight, then any information supporting that preconception will draw his or her notice and be recalled; however, any information not supporting that preconception will be ignored and forgotten.

Taken together, short-term memory limitations, serial position effect, and confirmation bias can be seen to have a significant impact on how information should be presented to pilots:

- It is clear that pilots should not be expected to hold large amounts of information in their short-term memory. Information presented early in a briefing may have been forgotten or displaced by information presented later. Thus, asking pilots to draw conclusions and make judgments based on comparisons or combinations of data presented over the course of an extended weather briefing is unrealistic. At a minimum, these data must be chunked, or combined in meaningful ways, so as to reduce the memory burden on the pilot.
- Information particularly germane to the safety of a flight should be placed at the beginning or end (preferably, both) of the briefing to maximize recall.
- Both the weather briefer and pilots need to be aware of the tendency to attend selectively to information that confirms pre-existing concepts. Because the weather briefer usually adheres to a standard format, there is less likelihood that he or she will selectively brief the pilot based on the briefer's biases, but there has been no research that demonstrates that effect. However, the effect of confirmation bias on the receiver of information is well established, and only adherence to a disciplined approach to flight planning will allow the pilot to overcome this tendency.

3.8 CURRENT ISSUES

The issues currently facing aviation psychology and human factors are an outgrowth of the adoption of the glass cockpit design for air transport aircraft over two decades ago, as well as the gradual infiltration of computers and computer-based technology into the flight decks and air traffic control systems during that period. The introduction of new flight control and management systems has not been without its own set of difficulties. Although the FMS takes over many of the tasks previously performed by crew members, it introduces new tasks or its own. These tasks, which might be referred to globally as "managing the management system," involve more planning and problem solving in place of the old psychomotor tasks now performed by the FMS.

The design of the human interface to these systems has not been entirely satisfactory, and issues over mode confusion still arise. The design of a system in which the human is kept constantly aware of the present state and future state of the system has not yet been entirely achieved. This problem is likely to grow even more acute as FMS-like systems are installed in growing numbers of general aviation aircraft, where they may be operated by pilots with very limited experience and training. A completely intuitive design for these systems will become a necessity because the training required to operate the current systems will not be feasible.

This lack of an intuitive design is evident in the current generation of GPS navigation systems for light aircraft. Although GPS allows for highly accurate three-dimensional navigation at almost any point on the surface of the Earth, its implementation has been subject to a great deal of criticism. Almost without exception, the displays used for GPS in light aircraft are small and the controls crowded together. Further, the functions of the controls are mode dependent, and the execution of tasks that in the traditional VOR navigation system required only a few tasks now requires extensive scrolling through menus and multiple function selections. For example, a VOR approach requires approximately five discrete steps, while the corresponding GPS approach requires over a dozen.

Human factors issues associated with GPS displays and controls have been the subject of extensive research in recent years. Leading these efforts have been the research arms of the aviation regulatory agencies, particularly in the United States, Canada, and New Zealand. Researchers at the Civil Aerospace Medical Institute of the Federal Aviation Administration have conducted several studies to identify problems in the usability of GPS receivers. These studies have ranged from evaluations of individual receivers to more wide-ranging global assessments and have identified many shortcomings of the GPS receiver interface.

Wreggit and Marsh (1998) examined one specific unit that was believed to be typical of a class of devices available at the time. They had nine general aviation pilots perform 37 GPS-related tasks requiring waypoint setting, GPS navigation, and GPS data entry and retrieval. Their results indicated that a number of the menu structures used by the device interfered with the pilots' successful entry of data, editing of stored data, and activation of functions. Based on their findings, Wriggit and Marsh provided recommendations for the redesign of the interface structure. Some of the specific recommendations included consistent assignment of a given function to one button, provision of consistent and meaningful feedback, and provision for an "undo" or "back" function that would reduce the number of button presses.

Williams (1999a, 1999b) conducted an extensive review of user interface problems with GPS receivers, using data collected from interviews with subject matter experts in the Federal Aviation Administration and from an inspection of the observation logs from an operational test of a GPS wide area augmentation system. Although he notes several interface issues associated with the displays and controls, Williams (1999a, p. 1) notes that "probably the most significant feature of GPS units, as far as the potential for user errors is concerned, is the sheer complexity involved in their operation." He points out that one measure of this complexity is the size of the instruction manual that accompanies each unit. The operation of the radio and display for a traditional VOR navigation system could be explained in, at most, 10 pages; however, manuals for GPS receivers typically contain 100–300 pages. Although there does not seem to be any published research on the subject, one can wonder just how many pilots have actually read all the instructions that accompany their GPS receiver. One might also speculate on how much of that material is actually retained.*

In addition to the overriding issue of complexity, Williams (1999a, 1999b) identifies a large number of specific human factor issues that detract from GPS receiver usability—many of which are hauntingly reminiscent of the problems identified by Fitts and Jones a half-century earlier. Some examples of the issues identified by Williams include:

Button placement. Inadvertent activation of the GPS buttons, just like inadvertent activation of landing gear and flaps, is made more likely by the poorly considered placement of the buttons. In the example provided by Williams, a manufacturer has elected to place the "clear" button between the "direct-to" and the "enter" buttons. This is an unfortunate arrangement because activation of the "direct-to" button is normally followed by the "enter" button. Placement of the "clear" button between these two buttons makes it much more likely that the pilot will activate the "clear" button, when the intention was to activate the "enter" button. Recovery from such an error may entail considerable reprogramming of the GPS, perhaps at a time when the pilot is experiencing high workload from other activities, such as executing a missed approach.

Knob issues. Many GPS receivers use a rotary knob to select and enter information. Often these knobs are used to select the alphanumeric characters of airports, VORs, and other navigation waypoints. Some of the knobs do not allow users to backtrack, so if they overshoot the character they wanted, they must continue turning until they go through the entire list again. This can significantly increase the head-down time required to program the receiver—a problem that is particularly acute while in flight. Furthermore, these knobs may function in more than one physical position: either pulled out or pushed in; the two positions provide entirely different functionality. Because there is no signal, other than a faint tactile sensation, to indicate which mode the knob is in, pilots can only determine what the knob is going to do by turning it and observing what happens. Clearly, this is an arrangement ripe with potential for serious errors, particularly when the pilot's attention is directed elsewhere.

---

* In addition to printed manuals, manufacturers also offer online tutorials on the principles of GPS navigation. How much this helps with the actual use of their systems during flight is debatable. Trimble Navigation: http://www.trimble.com/gps/index.shtml+Garmin Navigation: http://www8.garmin.com/aboutGPS/

Button labels. Williams (1999a, p. 4) notes that "buttons that perform the same type of task on different units can have different labels." The lack of uniformity, coupled with the complexity mentioned earlier, makes it difficult for a pilot familiar with one GPS system to use a different system.

Automatic versus manual waypoint sequencing. Readers of pilot reports captured by the Aviation Safety Reporting System (ASRS)* soon come to recognize the familiar question posed by pilots of aircraft with modern FMSs. That question is typically "What is it doing?" Alternatively, the question may be stated as "Why is it doing that?" The "it" in both questions is the FMS and/or autopilot, and the questions are raised because the pilot is suffering from what is commonly termed mode confusion. The aircraft is behaving in a way that is not consistent with the pilot's mental model of what it should be doing. This discrepancy arises because the complexity of the FMS allows for it to operate in multiple modes. If the pilot thinks that the FMS is in one mode, but it is actually in another, then truly unexpected things can happen, occasionally resulting in accidents. One such example is the Air France Airbus A-320 that crashed in Mulhouse-Habsheim Airport, France, following a low altitude fly-by (Degani, Shafto, and Kirlk 1996).

Regrettably, pilots of general aviation aircraft, who are often envious of the equipment and capabilities of transport category aircraft, now have the dubious honor of sharing the problem of mode confusion with their airline transport brethren. Williams (1999a, p. 4) reports that "one of the most often cited problems…involved either placing the receiver in a mode where it automatically sequences from one waypoint to the next during the approach or in a nonsequencing mode." Winter and Jackson (1996) reported that pilots frequently forgot to take the GPS receiver out of the "hold" function after completing the procedure turn. Because of that error, they were unable to proceed to the next approach fix.

Many of these deficiencies were also noted by Adams, Hwoschinsky, and Adams (2001) in their review of adverse events attributed to GPS usage. In addition to the usability issues identified by Williams (1999a, 1999b), such as button placement and display size, Adams et al. also highlighted problems that arose from pilot overreliance on GPS, programming errors, and lack of knowledge on the use of the GPS receivers. As a further illustration of the relative complexity of the GPS receivers, Adams and colleagues noted that although 5 steps were necessary to perform an approach using the traditional VOR system, 13 steps were required in an equivalent GPS approach.

Earlier, we noted that the traditional approach to the development of highly usable systems was based on the sequence: design, guard, warn, and train. Although GPS navigation arguably represents a remarkable step forward in the technology of air (and surface) transportation, GPS receivers (at least those marketed for general aviation aircraft) represent an equally remarkable failure to adhere to this philosophy. A failure to design usable receivers that prevent or mitigate errors leads to the necessity to warn users about their shortcomings and to attempt to remediate the problems by

---

* http://asrs.arc.nasa.gov/main_nf.htm

training users so that they do not fall victim to the interface idiosyncrasies. It is sad to note that, 60 years after Fitts and Jones showed us how errors could be prevented by a focus on the user and simple design changes, many designers (and regulators) still have not taken their lessons to heart. It is a situation that is cogently described by Dekker (2001) in a report that he aptly titled, "Disinheriting Fitts & Jones '47."

3.9 SUMMARY

The central message that the reader should take away from this chapter is that systems—mechanical systems, social systems, training systems, display systems— must be designed so that they conform to the characteristics of their users and the tasks that they must perform. The design of systems is an engineering process marked by a series of trade-off decisions. The designer may trade weight for speed, increased power for increased reliability, or the size and legibility of displays for the presentation of additional information. The list is almost endless. In each of these design decisions, the engineer is striving to meet some design criteria, without being able to meet all criteria simultaneously equally well. In our everyday world, we often wish to satisfy competing criteria. For example, we might wish to have a very large house and simultaneously wish to have a very small monthly house payment. Unless a rich uncle dies and leaves us a pot of money, we are forced to compromise with a house that is big enough and a mortgage payment that is not too big.

Usually, engineers produce a workable design that does not sacrifice the elements critical to successful operation of the system for the sake of competing criteria. Sometimes, however, they produce controls with the same knobs (it saves production costs to have all the knobs identical), leading to confusion during moments of high workload. They may also produce instrument panels in which essential, if rarely used, information is hidden physically, on a dial that cannot be seen without a great deal of effort, or logically, as part of a multifunction display system in which the needed information lurks beneath two or three levels of menus. However, the reasons that systems are poorly designed from the standpoint of the human user do not serve as excuses for those designs.

The reader should now be aware of some of the features and considerations that go into the production of usable aviation systems. We hope that readers will use that knowledge at the least to be informed consumers of those systems and, even more, to become active advocates for improved aviation systems.

RECOMMENDED READING

Diaper, D., and Stanton, N. 2004. *The handbook of task analysis for human–computer interaction*. Mahwah, NJ: Lawrence Erlbaum Associates.
Endsley, M. R. 2003. *Designing for situation awareness: An approach to user-centered design*. New York: Taylor & Francis.
Lidwell, W., Holden, K., and Butler, J. 2003. *Universal principles of design: 100 ways to enhance usability, influence perception, increase appeal, make better design decisions, and teach through design*. Gloucester, MA: Rockport Publishers.

Noyes, J. 1999. *User-centered design of systems.* New York: Springer–Verlag.

Salvendy, G. 1997. *Handbook of human factors and ergonomics,* 2nd ed. New York: Wiley.

Stanton, N. 2005. *Handbook of human factors and ergonomics methods.* Boca Raton, FL: CRC Press.

Stanton, N. A., Salmon, P. M., Walker, G. H., Barber, C., and Jenkins, D. P., eds. 2005. *Human factors methods: A practical guide for engineering and design.* Aldershot, England: Ashgate Publishing Ltd.

Woodson, W. E. 1992. *Human factors design handbook.* New York: McGraw–Hill.

REFERENCES

Adams, C. A., Hwoschinsky, P. V., and Adams, R. J. 2001. Analysis of adverse events in identifying GPS human factors issues. In *Proceedings of the 11th International Symposium on Aviation Psychology,* Columbus: Ohio State University.

Aircraft Owners and Pilots Association 2006. *2006 Nall report: Accident trends and factors for 2005.* Frederick, MD: Author.

Annett, J., and Stanton, N. 2000. *Task analysis.* New York: Taylor & Francis.

Boff, K. R., Kaufman, L., and Thomas, J. P., Eds. 1988. *Handbook of perception and human performance.* New York: John Wiley & Sons.

Booher, H. R. 1990. *Manprint.* New York: Springer.

———. 2003. *Handbook of human systems integration.* New York: Wiley.

Ciavarelli, A., Figlock, R., Sengupta, K., and Roberts, K. 2001. Assessing organizational safety risk using questionnaire survey methods. In *Proceedings of the 11th International Symposium on Aviation Psychology.* Columbus: Ohio State University.

Cohen, D., Otankeno, S., Previc, F. H., and Ercoline, W. R. 2001. Effect of "inside-out" and "outside-in" attitude displays on off-axis tracking in pilots and nonpilots. *Aviation, Space, and Environmental Medicine* 72:170–176.

Degani, A., Shafto, M., and Kirlk, A. 1996. Modes in automated cockpits: Problems, data analysis, and a modeling framework. *Proceedings of the 36th Israel Annual Conference on Aerospace Sciences.* Haifa, Israel.

Dekker, S. W. A. 2001. Disinheriting Fitts & Jones '47. *International Journal of Aviation Research and Development* 1:7–18.

———. 2002. The re-invention of human error (technical report 2002-01). Lund University School of Aviation. Ljungbyhed, Sweden.

———. 2003. Punishing people or learning from failure? The choice is ours. Lund University School of Aviation. Ljungbyhed, Sweden.

Fitts, P. M., and Jones, R. E. 1947a. Analysis of factors contributing to 460 "pilot error" experiences in operating aircraft controls. Memorandum report TSEAA-694-12, Aero Medical Laboratory, Air Materiel Command, Wright-Patterson Air Force Base, OH.

———. 1947b. Psychological aspects of instrument display. I: Analysis of 270 "pilot-error" experiences in reading and interpreting aircraft instruments. Memorandum report TSEAA-694-12A, Aero Medical Laboratory, Air Materiel Command, Wright-Patterson Air Force Base, OH.

———. 1961a. Analysis of factors contributing to 460 "pilot error" experiences in operating aircraft controls. In *Selected papers on human factors in the design and use of control systems,* ed. E. W. Sinaiko, 332–358. New York: Dover Publications.

———. 1961b. Psychological aspects of instrument display. I: Analysis of 270 "pilot-error" experiences in reading and interpreting aircraft instruments. In *Selected papers on human factors in the design and use of control systems,* ed. E. W. Sinaiko, 359–396. New York: Dover Publications.

Helmreich, R. L., Merritt, A. C., and Wilhelm, J. A. 1999. The evolution of crew resource management training in commercial aviation. *International Journal of Aviation Psychology* 9:19–32.

Hunter, D. R. 1995. Airman research questionnaire: Methodology and overall results (DOT/FAA/AAM-95/27). Washington, D.C.: Federal Aviation Administration, Office of Aviation Medicine.

Kirwan, B., and Ainsworth, L. K. 1992. *A guide to task analysis.* London: Taylor & Francis.

Meister, D. 1985. *Behavioral analysis and measurement methods.* New York: John Wiley & Sons.

Miller, G. A. 1956. The magical number seven, plus or minus two: Some limits on our capacity for processing information. *Psychological Review* 63:81–97.

Previc, F. H., and Ercoline, W. R. 1999. The "outside–in" attitude display concept revisited. *International Journal of Aviation Psychology* 9:377–401.

Sanders, M. S., and McCormick, E. J. 1993. *Human factors in engineering and design,* 7th ed. New York: McGraw–Hill.

Seamster, T. L., Redding, R. E., and Kaempf, G. L. 1997. *Applied cognitive task analysis in aviation.* Aldershot, England: Ashgate Publishing Ltd.

Shepherd, A. 2001. *Hierarchical task analysis.* New York: Taylor & Francis.

Sinaiko, H. W., and Buckley, E. P. 1957. Human factors in the design of systems (NRL report 4996). Washington, D.C.: Naval Research Laboratory.

———. 1961. Human factors in the design of systems. In *Selected papers on human factors in the design and use of control systems,* ed. E. W. Sinaiko, 1–41. New York: Dover Publications.

Wickens, C. D. 2003. Aviation displays. In *Principles and practice of aviation psychology,* ed. P. S. Tsang, and M. A. Vidulich, 147–200. Mahwah, NJ: Lawrence Erlbaum Associates.

Wickens, C. D., and Hollands, J. G. 2000. *Engineering psychology and human performance,* 3rd ed. Upper Saddle River, NJ: Prentice Hall.

Williams, K. W. 1999a. GPS user-interface design problems (DOT/FAA/AM-99/13). Washington, D.C.: Federal Aviation Administration.

———. 1999b. GPS user-interface design problems: II (DOT/FAA/AM-99/26). Washington, D.C.: Federal Aviation Administration.

Winter, S., and Jackson, S. 1996. GPS issues (DOT/FAA/AFS-450). Oklahoma City, OK: Federal Aviation Administration, Standards Development Branch.

Wreggit, S. S., and Marsh, D. K. 1998. Cockpit integration of GPS: Initial assessment—Menu formats and procedures (DOT/FAA/AM-98/9). Washington, D.C.: Federal Aviation Administration.

Yeh, M., and Chandra, D. 2004. Issues in symbol design for electronic displays of navigation information. In *Proceedings of the 23rd DASC Conference.* Salt Lake City, UT.

4 Personnel Selection

4.1 INTRODUCTION

Highly skilled people are essential for the airlines to operate efficiently, safely, and with satisfied customers. In a military context, the organization will also have other objectives, but skilled workers are still as important. For the individual employee, it is important to have a job that is sufficiently challenging, where the individual is appreciated and rewarded in relation to how well he or she performs the job. To achieve this, it is important to have both a good selection system and an effective training program for candidates who have been selected. A successful selection process will lead to lower dropout rates during training and an increase in the number of students completing the program. In addition, a well-designed selection system will, in the long term, contribute to a more effective and resilient organization; however, this claim may be harder to document compared to lowered dropout rates. Ideally, the selection of personnel should be based on methods that we know work for this purpose. This means that empirical evidence shows that the methods can tell us something about how the person will be able to perform the job or training he or she seeks.

Most of the research on selection methods in aviation has addressed the selection of pilots and, in recent years, air traffic controllers (ATCOs) also. The selection of military pilots is often based on young people without any previous flying experience; however, selection for civilian airlines includes experienced pilots as well as people without any flying experience (ab initio). Most airlines probably prefer to hire experienced pilots and thus avoid a long and expensive period of training.

In most cases, the selection of pilots and air traffic controllers is a comprehensive step-by-step process, at least for ab initio selection. That is, it usually starts with a large number of applicants who are tested with a range of psychological tests. In addition, applicants must meet a number of formal requirements in the form of medical requirements, no police record, and, sometimes, earlier education (e.g., completed high school or college). These formal requirements, however, may vary from organization to organization. After initial testing, the best candidates proceed to further testing and an interview and often more extensive medical examinations. For both ATCO and pilot selection, usually less than 10% of the applicant population will be accepted into the training program.

When choosing methods for the selection process, it is important to start with a thorough review of the job in order to determine what skills, abilities, and qualities are important for the person to possess. Such a systematic review is called a job analysis, and it involves a detailed survey of the tasks involved in the job. There are several ways to do this—for example, by observing workers or interviewing them. In the literature, various techniques are described that can be used to obtain information about the work content and the capabilities or skills needed to perform the tasks.

4.2 JOB ANALYSIS

A job analysis consists mainly of two elements: a job description and a person specification. A job description is an account of the activities or tasks to be performed; a person specification lists the skills, expertise, knowledge, and other physical and personal qualities the person must have to perform these activities. Several methods can be utilized to carry out a job analysis, and a job analysis can have multiple purposes in addition to forming the basis for a selection process. Therefore, some of the job analysis methods emphasize outlining the tasks to be solved or the behaviors that need to be performed, while others focus on the qualities the person should have. A job analysis is therefore both the job requirements (output) and personnel requirements (input). If the purpose of a job analysis is to determine what should be measured as part of the selection process, more emphasis should be put on the person specification. Figure 4.1 provides an overview of these two perspectives.

One of the most well-known job analysis methods is called the critical incident technique, which was developed by Flanagan during World War II for mapping the job performance of fighter pilots (Flanagan 1954). The main purpose of this method is to identify behavioral examples of good and poor performance. For many tasks, most people would have managed to perform them, so they are of little interest; others will be more demanding and not everyone will be able to perform the tasks successfully (e.g., critical incidents). The method involves using experienced workers as informants, and they are asked to describe critical tasks and what characterizes both good and bad ways to solve the tasks. It is also important to examine the extent to which these experts agree on what the critical tasks are and the characteristics of good and poor work performance. The work proceeds with some sorting of task descriptions, where the purpose is to create categories. Many tasks may require good communication, and many tasks may require the ability to perform calculations.

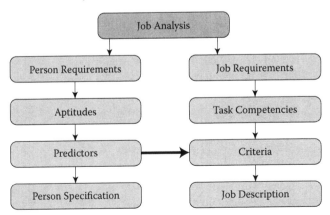

FIGURE 4.1 Two perspectives on job analysis.

Finally, a list is constructed of the abilities and skills the person must have to perform these tasks.

Another method, the repertory grid technique, was developed by George Kelly (1955). Here the experts are asked to imagine good, moderate, and poor workers. Then they are to describe how these workers are similar and different in relation to work performance. A third technique is Fleishman's job analysis survey method (1975). Employees are asked to indicate, on a seven-point Likert scale, the extent to which different abilities, personality traits, and skills are relevant to job performance. The person needs to consider seven main areas: cognitive abilities, psychomotor skills, physical demands, sensory capabilities, knowledge and skills, cooperation, and social skills. Each area has a number of subcategories. For example, cognitive ability includes a total of 21 categories, such as spatial orientation, time sharing, and attention.

4.2.1   JOB ANALYSIS FOR PILOTS AND AIR TRAFFIC CONTROLLERS

Fleishman's job-analysis method was used in a study of civilian pilots (Goeters, Maschke, and Eißfeldt 2004). Many of the cognitive abilities on the list were described as relevant or highly relevant, as were psychomotor and sensory abilities. Within the cooperative/social skills domain, coping with stress, communication, and decision making were identified as very important.

For military pilots, a revised version of the Fleishman method was used in a NATO study (Carretta, Rodgers, and Hansen 1996) in which pilots from several countries were asked to assess 12 critical tasks specific to the job of fighter pilot. They were then asked to specify the abilities and skills that were important when conducting these tasks. The most important abilities were situational awareness, memory, motivation, and reasoning. Least important were reading comprehension and writing, in addition to leadership.

Fleishman's method was also used in a study of German air traffic controllers (Eißfeldt and Heintz 2002). In addition to the original scales from Fleishman, a few more scales were added that included personal characteristics such as cooperation, communication, and the ability to handle stress. The cognitive abilities that received the highest ranking were speed of closure, visualization, and selective attention, in addition to time sharing. Few of the cognitive abilities received a low score. Visualization involves the ability to imagine objects and movements in space; selective attention means that an individual is able to concentrate on a task without being distracted. Time sharing involves the ability to shift attention quickly between different tasks. In addition, several psychomotor skills were rated as important, together with sensory abilities and specific knowledge (e.g., map reading). Several of the social skills were also highly rated, including stress resistance, decision making, and cooperation. There were some differences in the skills/qualities that were important between air traffic controllers in different functions (area control, approach, aerodrome control); however, for the most part, these differences were small.

4.2.2   A CRITICAL PERSPECTIVE ON JOB ANALYSIS

An important methodological question is to what extent we can rely on the results from a job analysis. One possibility is that experienced workers who are asked to assess the capabilities needed to perform the job may overestimate the number of skills and qualifications that are necessary. This may be more or less a conscious act, but it is natural that people want to present themselves in a favorable light, including overestimating the complexity of the job they are doing and the abilities and skills needed to perform the job.

In modern job analysis, more emphasis is placed on uncovering the competence needed and less emphasis is put on the specific tasks to be solved (see, for example, Bartram 2005). This is partly a function of the modern labor market, where many jobs are constantly changing and thus more global assessments and less focus on specific tasks and abilities may be more useful. However, it may be difficult to achieve a reliable assessment of the competence needed because competence is a complex concept often seen as a mixture of skills, knowledge, motivation, and interests. A meta-analysis of reliability coefficients from job analyses concluded that reviews of specific tasks had higher interrater reliability than the more general descriptions of the competence needed (Dierdorff and Wilson 2003).

The results from a job analysis may be used to select specific tests to be applied in the selection process and also to select appropriate criteria of work performance that could be used in a validation study. Many people would therefore argue that a job analysis is an important and necessary first step in a selection process. Meta-analyses have demonstrated, however, that ability tests predict job performance more or less independently of the occupation (Schmidt and Hunter 1998). One consequence of this may be that a very detailed job analysis may not be needed. On the other hand, job analyses of pilots and air traffic controllers have demonstrated that a number of highly specialized cognitive skills are important, and a test of general intelligence may not provide an adequate measure of such abilities.

4.3 PREDICTORS AND CRITERIA

The methods used to select applicants are identified as predictors, while measures of work performance are labeled criteria. When a psychological test is used to select an air traffic controller, the test is a predictor. To assess how well the test is suitable for this purpose, we have to conduct a validation study—that is, a study in which test results for applicants are compared to actual work performance or academic results. Both work performance and academic grades are examples of criteria.

4.3.1 PREDICTORS IN SELECTION

Predictors should ideally be selected because they measure something relevant for future work performance, possibly identified through a job analysis. A number of different methods can be used in a selection process, and here only the most common will be described. The *interview* as a selection method is applied to most professions, and it may be more or less structured. An interview is highly structured if the questions are formulated in advance and the ordering of the questions is also predetermined. Sometimes the interview is conducted toward the end of the selection process after the less time-consuming methods have been used. Employers in the process of hiring people probably also feel the need to meet the person face to face through an interview. Many employers believe that they have a unique ability to uncover who is more suited to the job and will fit nicely into the organization.

Unfortunately, this is assumption is often wrong. Unstructured interviews often have a very poor predictive validity, and the assumption that this method always identifies the right person is frequently wrong. More structured interviews, however, have a much higher predictive validity than those where random questions are asked. For an interview to be effective, it is important to think through and formulate job-relevant questions to be used with all the applicants. It is also important to train the interviewers, especially if more that one person is conducting the interviews. One advantage of the interview is that it also gives the applicant an opportunity to meet representatives of the organization and ask questions about the job.

Another type of predictor is the *assessment center,* which could be used for more purposes than just selection—for example, in leadership training and promotion. The assessment center method involves the candidate receiving various tasks that are similar or relevant to the job sought. Often, this involves situations in which small groups of people try to resolve a problem together. This makes it possible to study how people interact with each other, their leadership abilities, communication skills, and so on. Several trained observers, who usually make use of standardized forms to rate the performance, observe the applicants. The method is time consuming, and it often takes from half to a whole day or more.

Yet another type of predictor is called the *work sample test,* which represents a less comprehensive approach than an assessment center. This means that the candidate performs a similar task or the same task as the person would do as part of the job. The idea is that the behavior will predict similar behavior at a later date. There are often standardized scoring rules for how the performance should be rated.

A number of *psychological tests* also may be used as predictors, including ability tests and personality tests. Some of these are designed for selection, while others are designed for other purposes—for example, clinical use or diagnostics. Tests that are intended for special groups or designed for completely different purposes may not necessarily be suitable for personnel selection. Psychological tests are frequently used for pilot and air traffic controller selection, but less commonly for other groups in the aviation industry. Many of the tests used in the selection of pilots and air traffic controllers have been developed specifically for these occupational groups.

Past *work experience, school grades,* and *biographical data* are also sometimes used for selection purposes. If a person already has some work experience, it would be reasonable to obtain references from former employers. School grades may also be used in the selection process—in particular, to select people to further education. Biographical data gathered from employment records may also be used to discriminate between successful and unsuccessful employees. This information may be used for future selection of candidates. For example, if the best insurance sellers are married and own their own homes, then applicants who have these characteristics should be preferred. The items and their weighting are based on a purely empirical approach. Some people may argue that the method is unfair because applicants are selected based on factors over which they have little or no control—instead of measuring the relevant abilities and skills directly.

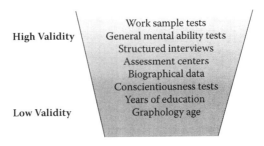

FIGURE 4.2 The validity of different selection methods. (Based on Schmidt, F. L., and Hunter, J. E. 1998. *Psychological Bulletin* 124:262–274.)

The final and perhaps most exotic method that will be mentioned here is graphology or handwriting analysis. This method involves an analysis of a person's handwriting in order to determine personality characteristics. Several studies have shown that this method is not suitable for selection, even though it is currently used in several European countries for personnel selection decisions (for an overview, see Cook 1998).

The various methods mentioned here have different predictive validity, and they are also different in relation to how costly and time consuming they are. Meta-analysis methods have been used to investigate the various methods' predictive validity, and the best predictors are work-sample tests, intelligence tests, and structured interviews, with an average predictive validity of about .50. Age and graphology had no predictive validity (Schmidt and Hunter 1998). An overview of the validity of different methods is presented in Figure 4.2.

4.3.2 CRITERIA OF JOB PERFORMANCE

Valid criteria for work performance are as important as good predictors when evaluating the selection process. The selection of criteria can be based on a previous job analysis where key tasks and what constitutes good performance have been identified. The easiest way to examine the predictive validity of a method is to use an overall criterion. For many jobs, one can argue that this is not adequate and that it would be more reasonable to apply multiple criteria to describe job performance and the variety of tasks performed. When conducting a validation study, the researcher needs to choose which criteria to use or, more conveniently, combined criteria instead of trying to predict a large number of criteria at the same time, which could be differentially related to the predictors.

In many cases, instructors or superiors are used to assess performance, and it is important that this happens in a systematic and reliable way. One way to assess reliability would be to let two instructors evaluate the same people and then study the degree of agreement between them. There are a number of known errors in such person evaluations; for example, if a person is good at one thing, it is automatically assumed that he or she performs other tasks equally well. It may also be difficult to get the observers to use the entire scale; that is, all performance may be assessed to be average, or there may be little variation between different items rated for each person. It is important that there is variation between individuals and between various tasks that the same person performs whenever the rating will be used as a criterion. If all the candidates perform the task equally well, it is not suitable as a criterion of work performance.

In order to achieve good interrater reliability it is important to train the observers and to specify what constitutes good and poor work performance. In addition, the criterion needs to have good construct validity, which means that it measures the construct in which one is interested—for example, leadership or communication skills. There will frequently be practical limitations to which criteria can be assessed as part of a validation study, and it is also important that these criteria are seen as relevant to the organization.

In relation to pilot selection, most of the criteria are usually obtained during training; in many cases, pass/fail in training is used. Pass/fail is a criterion about which it is easy to collect information that the organization regards as important. One problem with this criterion is that it is a somewhat indirect measure of performance. In some cases, reasons other than poor performance may cause the candidate to fail training, including everything from airsickness to lack of motivation. Nevertheless, in most cases, the criterion of pass/fail seems to work reasonably well. In a study of students at a Norwegian military flight school, the pass/fail criterion was highly correlated with assessments made by flight instructors (Martinussen and Torjussen 2004). In this study, pass/fail was taken as a valid measure of pilot performance.

Another problem with applying the criterion pass/fail is that it is based on performance during training and not actual work performance. However, few studies employ more long-term criteria of pilot performance, and there may be many reasons for this choice. A more long-term criterion would require that the validation study takes longer time to conduct. It may also be difficult to find comparable criteria for different jobs—for example, pilots working in different airlines. In addition, it is obviously more difficult to evaluate workers in a real-life setting than during training, where they expect to be evaluated.

In addition to pass/fail and instructors' ratings, assessments of graduates' performance in a simulator have also been used in validation studies. For air traffic controllers, the situation is similar, and validation studies have largely been conducted using criteria obtained during training or in a simulator.

4.4   HOW CAN WE KNOW THAT PREDICTORS WORK?

In order to document that predictors are useful in the selection of candidates, one can perform a local validation study or evaluate meta-analysis results that summarize relevant validation studies. Local validation studies are so named because they are conducted in the native setting (company, applicant, test, and training) in which the selection system under evaluation would eventually be employed. Local validation studies are usually performed by correlating test results with a measure of job performance—for example, performance in a simulator or assessments made by an instructor or supervisor. Sometimes, several tests are used in combination or tests are combined with an interview. In such cases, one can apply a combined test score or use regression analysis to find a weighted combination of predictors that gives the highest correlation with the criterion. In many cases, it may be difficult to conduct local validation studies because the organization does not employ a sufficient number of people within a certain time period.

4.4.1 META-ANALYSIS

An alternative to conducting a local validation study is to combine previous studies in a meta-analysis. In order to merge results from multiple studies, the studies must all supply a common metric or measure of effect. Fortunately, most of the articles reporting results from validation studies include correlation coefficients, which are highly suitable for meta-analysis. Some studies, however, only report the results from multiple regression analyses, and these cannot be combined with correlations from other studies. The meta-analysis calculation requires that a standardized index (e.g., the Pearson correlation coefficient or another measure of effect size) be used and that the results be reported for each predictor separately. In a regression analysis, the results indicate how well the combined set of tests predicts a criterion; regression coefficients will depend not only on the correlation between the test and the criterion, but also on the intercorrelations between other predictors included in the equation. Because the individual contributions of the predictor measures cannot be separated, the regression coefficients cannot be used in meta-analyses.

There are several meta-analysis traditions, and the most widely used method within the work and organizational psychology was developed by John Hunter and Frank Schmidt in the late 1970s. Their method was initially designed to study how well test validity could be generalized across different settings. The method is therefore well suited to perform a meta-analysis of validation studies because it takes into consideration many of the methodological issues relevant in such studies. Hunter and Schmidt (2003) have described a number of factors or circumstances that may affect the size of the observed correlation or validity coefficient. These factors, or statistical artifacts, will influence the size of the correlation coefficients in various degrees from study to study.

Three such statistical sources of errors are lack of reliability, restriction of range, and use of a dichotomous criterion (e.g., pass/fail) instead of a continuous measure. The lower the score reliability is, the lower the observed correlation will be. It is possible to correct for this artifact if the test score reliability is reported in the article (Hunter and Schmidt 2003):

$$r_{cor} = \frac{r_{obs}}{\sqrt{r_{xx}}\sqrt{r_{yy}}}$$

In this equation, r_{cor} is the corrected correlation, r_{obs} is the observed correlation, and r_{xx} and r_{yy} are reliability of the predictor and the criterion, respectively. The corrected correlation is an estimate of the correlation that would have been observed if the variables had been measured with perfect reliability. In some cases, it is appropriate to correct for lack of reliability in only one of the variables (e.g., correction for criterion reliability in validation studies) because the purpose is to evaluate the usefulness of the tests with all the errors and shortcomings that they may have. The correction is then

$$r_{cor} = \frac{r_{obs}}{\sqrt{r_{yy}}}$$

If we assume that the observed correlation between the ability test and a criterion is .40 and that criterion reliability is .70, then the corrected correlation is $.40/\sqrt{.70}$ = .48. This number represents the correlation that would have been observed if the criterion had been perfectly measured. The lower the score reliability is, the greater will be the correction factor.

The second factor that affects the size of the correlation is reduced test score variation (restriction of range) on one or both variables as a result of selection. This occurs if only the relationship between test scores and subsequent performance of those who have been selected on the basis of the test results is studied. If only the best half of the applicant group has been selected, then it will be possible to collect criterion data only for this group. The calculated correlation will be much lower for this group than if we had studied the entire unselected group. The effect on the observed correlation can be dramatic, given that a very small group is selected.

This problem is illustrated in Figure 4.3, where the shaded part represents the selected group included in the study. Based on the plot in this figure, we can see that if we calculate the correlation for the selected group, then the correlation would have been much lower than if the calculation had been based on the entire applicant group. If we assume an even stricter selection and move the line (X_1) to the right, then the shaded field will be almost like a round ball, implying zero correlation.

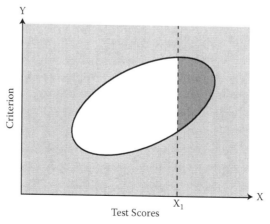

FIGURE 4.3 Illustration of restriction of range.

There are few empirical examples of the phenomenon "restriction of range." One of the few examples dates back to World War II, when applicants to the U.S. Air Force were tested and selected. Because of the lack of pilots at that time, all the applicants were admitted into the basic flying program. The predictive validity could then be calculated for the entire group as well as for a selected group. The predictive validity for the total test score (pilot stanine) was .64. If the normal procedure had been applied and only the top 13% of the candidates had been selected, then the predictive validity would have dropped to .18 (Thorndike 1949). This provides a picture of the dramatic effect that calculating the predictive validity on a highly selected group may have. In other words, the problem is that we have excluded the control group by only using the selected applicants in the study. The correction for range restriction is based on information about the test score standard deviation in the whole group, or the proportion of applicants selected (Hunter and Schmidt 2003). In situations where the selection is based on several tests or tests used in combination with other types of information, the situation becomes more complicated and more advanced models for range restriction correction should be applied (Lawley 1943; Johnson and Ree 1994).

The third statistical artifact is the application of a dichotomous criterion (e.g., pass/ fail in training) when the performance (flying skills) really can be said to be a continuous variable. This artifact also leads to a lower correlation between test and criterion than if we had measured performance on a continuous scale. This statistical artifact can be corrected if the distribution between the pass/fail ratios is known. The farther the distance is from a 50/50 distribution of pass/fail, the greater is the correction.

For both pilot and air traffic controller selection, all these statistical artifacts are frequently present and contribute to a lower observed correlation between the test and criterion. When possible, the observed correlations should therefore be corrected for these error sources before they are included in a meta-analysis, as well as to provide a better estimate of the true predictive validity. Unfortunately, such corrections are often difficult because the primary studies frequently lack the information needed to perform such corrections. Criterion reliability is rarely examined, and information about the selection ratio is not reported in many articles. The percentage passing or failing training is normally reported; this makes it possible to correct for the effect of using a dichotomous criterion. In studies where such corrections are not implemented, the observed correlations must be viewed as very conservative estimates of the tests' predictive validity. In addition, sampling error will contribute to variation between observed correlations in different studies. However, this error is unsystematic; therefore, the statistical methods described before cannot be used to correct for the error.

4.4.2 When Can Test Validity Be Generalized?

How do we know if the predictive validity of a test can be generalized over different settings? For example, can intelligence tests always be successfully used for selection, regardless of setting and occupation? Hunter and Schmidt (2003) proposed a rule of thumb that states that if at least 75% of the observed variance between the correlations can be attributed to statistical errors and sampling error, then it is reasonable to assume that the remaining variance is due to error sources not corrected for. In such instances, it is safe to assume that the true variance between studies is very small or zero and that the mean correlation is an appropriate estimate of the true validity.

The second situation arises when we have real variance in the population; it is then possible to estimate an interval (credibility interval) that, with a high probability, includes the predictive validity. This interval is calculated based on the corrected average correlation and the estimated population standard deviation (Whitener 1990). If the interval is large and in addition contains zero, it means that the actual variation between studies is considerable and that the test in some cases does not have predictive validity. Other occasions may arise, however, in which the interval is of a certain size, but does not include zero. This means that there is some variation in the predictive validity, but that it is always larger than zero. In that case, the predictive measures can be used as a valid predictor, even though its utility will vary from instance to instance.

It is also possible to perform a significance test of the variation between studies, but this is a less common strategy in the Hunter and Schmidt meta-analysis method, where the estimation of variance is emphasized.

4.5 HISTORICAL OVERVIEW

4.5.1 Pilot Selection

Probably few professions have been tested as much as pilots (see, for example, Hunter 1989). During World War I, the first tests were developed and validated—not many years after the Wright brothers made their first flight (Dockeray and Isaacs 1921). Many of the first tests were simple constructions simulating tasks or situations with which humans involved in flying would have to cope. One of the earliest test batteries from the United States (Henmon 1919) contained tests that measured emotional stability, reaction time, general cognitive abilities, and sense of equilibrium.

In Europe, similar tests were developed in several countries. In Denmark, Alfred Lehman (Termøhlen 1986) developed methods for pilot selection in his laboratory. He suggested tests that measured emotional stability, evaluation of spatial relationships, attention, reaction time for sound, and sense of equilibrium. The test that measured emotional stability consisted of psychophysiological measurements at the same time as a test administrator fired a shot behind the back of the candidate. Lehman suggested that the test was unsuitable for selection because it was not possible to distinguish those who were really cold blooded from those who reacted to stress induced in the test situation (Termøhlen 1986). There were many similarities between the tests that were used in different countries in this first phase of test development. Paper-and-pencil tests were used together with apparatuses that simulated aspects of a flying machine, in addition to simple measures of reaction time and judging distance and time.

After World War I had ended, there was little research on a pilot selection in most countries (Hilton and Dolgin 1991; Hunter 1989). An exception was Germany, where a large number of tests were developed; at the beginning of World War II, the country had a test battery that consisted of 29 tests that measured, among other things, general intelligence, perceptual abilities, coordination, ability, character, and leadership (Fitts 1946). During the war, this test battery was replaced by a less extensive system with fewer tests and more emphasis on references and interview data (Fitts 1946).

In England, the United States, and Canada, the trend was different. At the start of the war, few tests were in use; by the end of the war, a large number of tests had been developed and implemented. In Norway, the Norwegian Air Force used tests first in 1946 (Riis 1986). Since then, the Norwegian test battery has been expanded and validated several times (see, for example, Martinussen and Torjussen 1998, 2004; Torjussen and Hansen 1999).

After World War II, research declined again, and many countries put emphasis on maintenance rather than on developing new tests. This more or less continued until the first computerized tests were invented in the 1970s and 1980s (Bartram 1995; Hunter and Burke 1987; Kantor and Carretta 1988). As computer technology became cheaper and better, the paper-and-pencil tests were replaced entirely or partially with computerized tests in most Western countries (Burke et al. 1995).

Early in the history of aviation, personal qualities, as well as cognitive and psychomotor skills, were seen as important in order to become a good pilot. To examine which personality traits were important, both observation of pilots and participating observation were used. After having undergone flight training, Dockeray concluded that "quiet methodical men were among the best flyers, that is, the power and quick adjustment to a new situation and good judgment" (Dockeray and Isaacs 1921). It would still be many years before personality tests were developed and used for pilot selection. In the United States, a comprehensive program to find suitable personality measures for pilot selection was started in the 1950s. The research program was led by Saul Sells (1955, 1956), and a total of 26 personality measures were evaluated. Sells and his colleagues used more long-term criteria of pilot performance in their evaluation, and they concluded that personality tests were better predictors of long-term criteria compared to ability tests, where the predictive validity declined over time.

A number of well-known personality inventories have also been examined in relation to pilot selection over the years. These include the MMPI (Minnesota Multiphasic Personality Inventory) (Melton 1954), Eysenck Personality Inventory (Bartram and Dale 1982; Jessup and Jessup 1971), Rorschach (Moser 1981), and Cattell 16PF (Bartram 1995). The results showed only low to moderate correlations with the criterion. One of the few studies on civilian pilots (conducted at Cathay Pacific Airlines) showed that, based on training results, successful pilots scored lower on anxiety compared to less successful pilots (Bartram and Baxter 1996).

In Sweden, a projective test called the "defense mechanism test" (DMT) was developed by Ulf Kragh (1960). The purpose was to select applicants for high-risk occupations such as pilots and deep-sea divers. The test material consisted of pictures presented using a special slide projector that displayed images. Exposure of each picture was very short, but increased each time the image was presented. The person drew and explained what he or she saw, and the discrepancy between the actual image and what the person reported was then interpreted as various defense mechanisms. (This is a very simplified overview of a complicated and comprehensive scoring procedure; see, for example, Torjussen and Værnes 1991.) The test was met with considerable optimism when it was launched, and it was tested on military pilots in several countries such as England, The Netherlands, and Australia, as well as in Scandinavia (Martinussen and Torjussen 1993).

However, it has been difficult to document the predictive validity of the test for pilots outside the Scandinavian countries, and very few countries currently use the test. With the introduction of computers in testing, a number of personality-related concepts have been evaluated. This includes measures of risk taking, assertiveness, field dependency, and attitudes (see Hunter and Burke, 1995, for an overview). In most cases, this has only resulted in very small correlations with the criterion and no increase in the predictive validity beyond what the ability tests predicted (i.e., no incremental validity). Recent studies, however, have yielded more positive results for personality measures used for pilot selection (see, for example, Bartram and Baxter 1996; Hörmann and Maschke 1996).

Personality traits are emphasized today in varying degrees during the selection process. Some countries, like the United States and England, do not use personality tests in their selection process (Carretta and Ree 2003). However, an assessment of personal qualities and motivation may be evaluated during the interview.

In addition to ability, psychomotor and personality tests, biodata, past experience, and flying performance in the simulator have been used as predictors in varying degrees over the years.

4.5.2 SELECTION OF AIR TRAFFIC CONTROLLERS

Selection of applicants for air traffic controller education occurs in most Western countries by using psychological tests, but research in this area is less extensive compared to that for pilot selection (Edgar 2002). The first psychological tests were put into use early in the 1960s and consisted of paper-and-pencil tests (Hätting 1991). Today, computerized tests are used in many countries, and the selection process is often as comprehensive as that for pilots.

Most validation studies in relation to air traffic controllers have been conducted by the Federal Aviation Administration (FAA). The first test batteries adopted in the 1960s had paper-and-pencil tests measuring reasoning (verbal and numerical), perceptual speed, and spatial skills (Hätting 1991). Test results were combined with information about education, age, and experience in the selection process. In the 1970s, the FAA began developing a simulation-based test that would measure the candidates' skills in applying different rules in a simulated airspace. The test was later adapted to a paper-and-pencil format and labeled "multiplex controller aptitude test." It was used together with measures of reasoning ability and professional experience in the selection from the beginning of the 1980s. The development of a computerized test battery began in the 1990s, and it measured, among other things, spatial reasoning, short-term memory, sense of movement, pattern recognition, and attention (Broach and Manning 1997).

In Europe, EUROCONTROL conducted (Hätting 1991) a review of the member states' selection procedures in the late 1970s and discovered that most countries applied tests that measured spatial perception, verbal ability, reasoning, and memory. Few countries used tests to map out the interest or motivation for the profession. An exception was Germany where, in addition to a comprehensive test battery, the

German Aerospace Center (Deutsches Zentrum für Luft-und Raumfahrt) also used a measure of personality traits and a simulation-based test to measure cooperation (Eißfeldt 1991, 1998). In addition, all countries had medical requirements and formal requirements concerning age and previous education.

In the 1980s, computerized tests were developed in Germany and England, and in 2003 EUROCONTROL launched a common computer-based test battery (called FEAST) for the selection of air traffic controllers that the member countries could apply. As part of this project, the member countries would also have to supply data to a joint study of the predictive validity of the tests.

Today, many countries have replaced paper-and-pencil tests entirely or partially by computerized tests. Basic cognitive abilities are measured; in addition tasks are assigned in which computer technology is used to simulate parts of the work as an air traffic controller. In Sweden, a considerable effort has been put into developing a situational interview, where the purpose is to map individual abilities and social attitudes (Brehmer 2003). Situation interviews are developed from critical incidents, and the main goal is to determine effective versus ineffective work performance by asking the applicants very specific questions. Some personality tests have also been investigated, but the results have generally been discouraging, with weak correlations between measures and the criterion. Studies based on the big-five model, however, have found more positive results (Schroeder, Broach, and Young 1993).

4.6   HOW WELL DO THE DIFFERENT METHODS WORK?

Fewer validation studies have been conducted for air traffic controllers compared to pilots. In 2000, a meta-analysis of available studies found a total of 25 articles and reports that documented validation results for ATC selection based on a total of 35 different samples (Martinussen, Jenssen, and Joner 2000). These studies were published between 1952 and 1999, and the majority were based on applicants and students (92%). Most of the studies were conducted in the United States (77%), and the criteria used were mostly collected during training (e.g., pass/fail, instructor evaluations, simulators).

The results from these 25 articles were combined in a meta-analysis where the average predictive validity was calculated, and the population variance of the tests was estimated. The total samples ranged between 224 and 11,255 persons for the different categories of predictors. Virtually none of the studies reported information that made it possible to correct for lack of reliability in the criterion and restriction of range. The average correlations are therefore an underestimate of the true predictive validity. Correlations were corrected for the use of a dichotomous criterion (pass/fail). The various tests and predictors were grouped into categories, and a summary of these results is presented in Figure 4.4. For all but two of the predictors (verbal skills and multitasking), there was some true variance between studies.

When it comes to pilots, a large number of validation studies have been conducted since World War I. Several literature reviews (see, for example, Carretta and Ree 2003; Hunter 1989) and two meta-analyses have been published (Hunter and Burke 1994; Martinussen 1996). In spite of a slightly different database and some procedural differences in the way the meta-analyses were conducted, the two

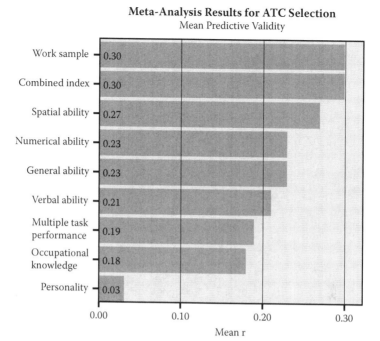

Meta-Analysis Results for ATC Selection
Mean Predictive Validity

FIGURE 4.4   Meta-analysis results for air traffic controllers.

meta-analyses resulted in very similar findings. An overview of the average correlations for the different test categories from Hunter and Burke (1994) is presented in Figure 4.5. A total of 68 studies published between 1940 and 1990 were included with a total sample of 437,258 participants. The average correlation was not corrected for any statistical artifacts because the primary studies did not include the necessary information to make this possible. For all test categories, there was some true variation between studies, and for some test categories the credibility interval included zero, implying that the predictive validity in some situations was zero. This applied to the categories general intelligence, verbal skills, fine motor ability, age, education, and personality. This means that, for the other categories, the predictive validity was greater than zero even though there was some true variance between studies.

4.7   PERSONALITY AND JOB PERFORMANCE

Meta-analysis results (Martinussen 1996; Martinussen et al. 2000) for both pilots and air traffic controllers have shown that cognitive ability tests can be used effectively in the selection process, but the results are less impressive when it comes to personality measures. This finding may be due to several factors and does not imply that personality is not important for work performance. On the contrary, job analyses for both pilots and air traffic controllers have listed personality characteristics as important in order to do a good job. For air traffic controllers, cooperation, good

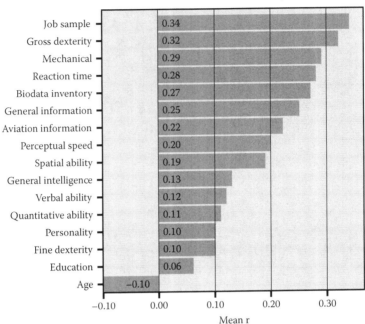

FIGURE 4.5   Meta-analysis results for pilots.

communication skills, and ability to cope with stress were emphasized (Eißfeldt and Heintz 2002). For military pilots, qualities such as achievement motivation and ability to make decisions and act quickly, in addition to emotional stability, have been seen as particularly important (Carretta et al. 1996). In another study, American fighter pilots ($N = 100$) were asked to rate 60 personality traits in relation to various aspects of the job. The most important of these dimensions was conscientiousness (Siem and Murray 1994).

How can we then explain the lack of predictive validity? One possibility is that the personality tests used have not been suitable for selection purposes. For example, clinical instruments originally designed to diagnose problems or pathology may not be appropriate for personnel selection. Another possibility is that many of these tests are self-reported; applicants choose the more socially desirable response and thus it is easy for applicants to present themselves in a favorable light. A meta-analysis of personality measures based on the big-five model showed that although the applicants to some degree presented themselves in a favorable light, this had little effect on the predictive validity of the measures (Ones, Viswesvaran, and Reiss 1996).

Another factor is that the criteria used are often obtained during training, and it is reasonable that some cognitive abilities are more important in this educational setting compared to personality traits. This is in line with the findings of Sells (1955, 1956) suggesting that personality tests were better predictors in the longer term.

However, few validation studies have used actual job performance providing empirical evidence for this statement.

In a study of 1,301 U.S. pilot students, the results indicated that the male candidates were more outgoing and scored lower on agreeableness compared to the norm. When examining the subscales of the big-five inventory, several differences between pilots compared to the normal population can be found: They are less vulnerable (neuroticism); they are active, outgoing, and seek new experiences (openness); and they are coping oriented and competent (conscientiousness). Female pilots showed many of the same characteristics compared to a normative sample of women; in addition, they scored higher on openness to new experience. In other words, they like to try out new things (Callister, King, and Retzlaff 1999).

In another study of 112 pilots from the U.S. Air Force, female pilots were compared with male pilots and with a random sample of women. Female pilots scored higher than their male colleagues on the dimensions agreeableness, extroversion, and conscientiousness. They were also more emotionally stable and scored higher on openness (King, Retzlaff, and McGlohn 1998).

A recent study of U.S. pilot students showed that, compared with normative data, these pilots could be described as people who set themselves high goals and engaged in constructive activities to achieve these goals. The goals often included new and unfamiliar experiences and also the quest for increased status, knowledge, and skills. They appeared often as calm, less inhibited, and more willing to tolerate risk compared to a normative sample (Lambirth et al. 2003).

In studies that have compared air traffic controllers with other professions, it has been found that air traffic controller students scored lower than other students on anxiety (Nye and Collins 1993). In another study using the big-five taxonomy, they scored higher on openness to new experience and conscientiousness and lower on neuroticism that a norm group (Schroeder et al. 1993).

There is, however, little evidence to support the notion of a fixed pilot personality or air traffic controller personality. Pilots and air traffic controllers vary on a number of personality characteristics in the same way that people in the general population vary. At the same time, there are some differences between the pilots and air traffic controllers as a group and the general population, probably as a function of the selection process and self-selection before people enter these professions.

4.8 COMPUTER-BASED TESTING

The first computerized tests were developed in the 1970s and 1980s (Bartram 1995; Hunter and Burke 1987; Kantor and Carretta 1988). As computer technology has become both cheaper and more efficient, paper-and-pencil tests have entirely or partially been replaced with computerized tests in most Western countries (Burke et al. 1995). The introduction of computerized testing has led to simplifications in test administration and scoring. But both computers and software need to be updated, so this type of testing also requires maintenance and revisions.

An advantage of the use of computerized testing is that it has made it possible to test more complex psychological abilities and skills than before. It is now possible to measure reaction time and attention, both alone and as part of a more complex task.

It is also possible to simulate parts of future work tasks and to present information both on the screen and through headphones. A problem associated with such complex dynamic testing is that the test may progress differently for different applicants. Options and priorities taken at an early stage could have consequences for both the workload and complexity of the task later.

In addition, it may be that applicants use different skills and strategies for problem solving. For example, someone may give priority to speed rather than safety and accuracy. This makes the scoring of such tests more complicated than with simpler tests, where the number of correct answers will usually be sufficient indices of performance. Another aspect of such dynamic tests is that they often require longer instruction and introduction periods before the testing can begin. This makes them time consuming. They are therefore often used at a later stage in the selection process, when the applicant group has already been tested with simpler tests and only the strongest candidates are permitted to enter the final phase of the selection process.

Computers have made it possible to apply more adaptive testing. That is, the degree of difficulty of the tasks is determined by how the candidate performs the first tasks in the test. This type of test is based on item response theory (see, for example, Embretson and Reise 2000) in which the purpose is to estimate the person's ability level. These tests are expensive in the developmental stage, and most tests used today are based on classical test theory.

With the Internet, it is now possible to test applicants located anywhere in the world. This is financially beneficial to the organization because it saves travel expenses. One problem with this approach, however, is how to ensure that the applicant is answering the test questions rather than someone else. Nevertheless, it is likely that the use of the Internet to conduct such pretesting will be adopted by more organizations. This allows applicants to investigate whether this is something for them to pursue. In the next step in the selection process, the best applicants will be invited to participate in further testing and whether the correct person actually took the tests can be checked. Some organizations provide applicants with information about the tests, and they are also given the opportunity to practice some of the tests before the selection process begins.

4.9 THE UTILITY OF SELECTION METHODS

Utility refers to how much an organization earns by applying specific methods in the selection process rather than using a more random selection process (for an overview of the topic, see Hunter 2004). Several models may be used in order to estimate this, and one critical factor in these models is, of course, the value of a good employee relative to a less efficient employee. A rule of thumb that has proven to hold for many occupations is that the best employees produce about twice as much as the worst. In utility calculations, the expenses associated with the selection process and testing have to be included, but this amount is often much smaller compared to that for a poor performing employee or a candidate who fails to complete an expensive training program. In addition, it is to be hoped that increased safety is also an outcome, but this is harder to document empirically because serious accidents are rare in aviation and those who could constitute the control group are normally not hired.

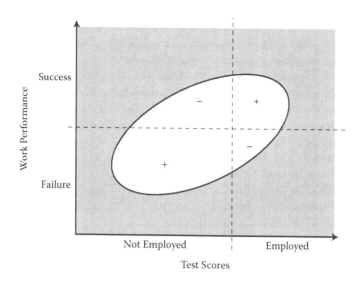

FIGURE 4.6   Right and wrong selection decisions.

Calculations performed by the German Aerospace Center in 2000 showed that the selection of ab initio candidates cost €3,900 per candidate, while the training cost €120,000. If the candidate fails training, the expenditure is estimated to be €50,000 (Goeters and Maschke 2002). The corresponding figure from the U.S. Air Force is between $50,000 and $80,000 for candidates who fail training (Hunter and Burke 1995). In other words, the test costs are relatively low compared to the costs for those who do not complete pilot training.

A simple model that may also be used to calculate the utility of the selection procedure is based on the Taylor and Russel (1939) tables. Their model assumes a dichotomous criterion, which is illustrated in Figure 4.6. According to the model, two correct decisions can be made:

1. Select those who would perform the job successfully.
2. Do not hire those who would not perform the job satisfactorily.

Two erroneous decisions can also be made:

1. Employ those who will not be successful (candidates who clearly pose the greatest problem for the organization).
2. Reject those who would manage the job.

The correct decisions are marked with (+) in the figure and the erroneous are marked with (−). The larger the correlation (in other words, the predictive validity) is, the more often correct decisions are made. To calculate the increase in correct decisions by using a given selection method, we need to know the predictive validity, the selection ratio, and how many would perform the job successfully in the applicant group (base rate).

For example, assume that the aviation authorities want to hire 20 people to perform security checks at a smaller airport. Assume that there are 100 applicants and that approximately 50% would be able to perform the job. If we do not use any form of selection process, but rather pick randomly, approximately half of the candidates would do the job satisfactorily. Suppose that we also use an ability test and the predictive validity is .40. To sum up, the selection rate is 20/100 = 0.20, base rate is 0.50, and the predictive validity is .40. What is the utility of using the test? By inspecting the table, we find that we increase the number that would succeed in the job from 50 to 73%. If the predictive validity is lower—for example, .20—the increase would be from 50 to 61%. If the selection rate is lower, fewer people are selected (e.g., only the top 5% of the applicants), and the proportion would increase from 50 to 82% (with r = .40). In other words, the lower the selection rate and the higher the validity are, the more advantage is provided by using the test.

The effect of changes in the base rate is somewhat more complicated, and the farther away from a base rate of 0.50, the smaller the increase in successful applicants is. Imagine a rather extreme situation (e.g., with no qualified applicants); it does not matter how high the predictive validity is because it will not improve the outcome; that is, no one would be able to do the job. Additional examples are presented in Table 4.1.

More sophisticated models for calculating the utility of selection tests do not make the assumption that performance is twofold (success/failure), but rather that it can be assessed on a more continuous scale. The most difficult part in these calculations is to estimate the dollar value of different workers. One way to do this is to assume a dollar value for a very productive employee—for example, by estimating how much it would cost the company to hire someone to perform the same job. A good employee means one who is in the top layer (i.e., the 85th percentile compared to a poor worker, who is at the 15th or lower percentile). In addition, the predictive validity and the quality of the applicants must also be known. These estimates can then be used to calculate how much the organization will earn per year per person employed. Included in these calculations are the costs of selection (see, for example, Cook 1998).

TABLE 4.1
Utility of Selection Methods

Base Rate	r	Selection Rate		
		0.10	0.30	0.50
10%	.10	.13	.12	.11
10%	.30	.22	.17	.14
10%	.50	.32	.22	.17
50%	.10	.57	.55	.53
50%	.30	.71	.64	.60
50%	.50	.84	.74	.67

Source: The examples are from Taylor, H. C., and Russel, J. T. 1939. *Journal of Applied Psychology* 32:565–578.

4.10   FAIRNESS IN SELECTION

Many countries have laws that prohibit discrimination on the basis of gender, race, and political or religious convictions when hiring. Some countries, such as the United States, also have laws against discrimination because of age or disability. Many countries have also adopted guidelines on the use of tests in employment where there are specified requirements for the test user and the methods. This is further discussed in Chapter 2.

A topic that has been discussed for a long time is how the concept of fairness should be understood. Is a method fair if it has the same predictive validity for different applicant groups, or is it only fair if the method will result in an equal number of people hired as they are represented in the applicant group or perhaps in the population? If, for example, women constitute 30% of the applicant group and a company only hires 5% of women applicants, then the selection method has adverse impact.

Opinions differ on which group should be used for comparison. Is it the total adult population in a country, or is it the group of qualified candidates that should be considered? The latter is probably the most reasonable choice when calculating the adverse impact. In the United States, the employer must ensure that the method used does not have a so-called adverse impact. The solution is then to be sure to hire a certain proportion of the groups that would have been underrepresented. Alternatively, the employer must argue that the test or method is job related and has predictive validity. That other equally valid methods do not have such a negative impact must also be ruled out. The legal practice on this topic appears to be far stricter in the United States (where many lawsuits have been instrumental in shaping today's practice) than in Europe.

4.10.1   DIFFERENTIAL PREDICTIVE VALIDITY

Two hypotheses are concerned with the claim that a test is valid for one group only (often white men), but not for others (e.g., different ethnic groups). One is that the test predicts job performance for one of the groups, but not for the other. The second hypothesis is that the tests are valid for both groups, but have a higher validity for one of the groups (differential validity). It has been difficult to document this, however, probably because the primary studies that have examined the predictive validity for the two groups have often included very small samples of minorities. Meta-analyses have not found support for the hypothesis of differential validity for white versus nonwhite, beyond what one might expect as a result of chance (Hunter, Schmidt, and Hunter 1979).

Similarly, in a meta-analysis where validity coefficients for women were compared with men, there was no overall support for differential test validity (Rothstein and McDaniel 1992). However, it turned out that for jobs that required little education and where there was a very high proportion of women or of men, some support for differential predictive validity was discovered. A study of U.S. Air Force applicants found no gender difference in tests' predictive validity in relation to flight performance (Carretta 1997).

All in all, little evidence supports the idea of differential validity, but this does not imply that all groups necessarily will have the same mean test score on all tests.

Fair methods imply that the test predicts job performance equally accurately in the different groups, rather than that groups have the same average test score.

4.11 APPLICANT REACTIONS

Most of the research in personnel selection has studied the development and validation of selection methods. It is easy to understand that an organization wants to focus on these areas in order to ensure that the methods with the highest predictive validity are used for selection. Another related perspective is to consider the selection process from the applicant's perspective. How does the person experience the selection methods and what kind of impression does he or she get from the organization on the basis of the selection process? The methods that applicants prefer are not necessarily the methods with the highest predictive validity.

Nevertheless, we should be concerned about the applicant's perspective for several reasons. First, applicants' perceptions of the methods' validity and fairness could influence their motivation to perform well on the tests. If the methods appear strange and unrelated to the job, applicants may be reluctant to do their best and perform well. This may also affect the chances that they will accept a future offer of employment from the organization. In a situation where young people have many opportunities for education and employment, it is important to attract the best applicants.

Several studies have examined what applicants or volunteers (often students) think about different methods, such as ability tests, personality tests, or interview. What applicants prefer depends to some extent on the context (i.e., the job sought) and whether they feel they have some control over what is going on. The content of the test or questions in the interview are also critical, and applicants in general prefer job-relevant questions.

Two studies of pilot applicants examined what the candidates thought about the selection process. One was based on the selection of pilots for Lufthansa (Maschke 2004), and the second study consisted of selection of pilots for the Air Force in Norway (Lang-Ree and Martinussen 2006). In both cases, the applicants were satisfied with the selection process and the procedures were considered to be fair. The applicants were also asked to rate different tests and methods used, and the computerized tests received the highest ratings. In the survey, the relationship between applicants' responses and test performances was also examined (Lang-Ree and Martinussen 2006). The results showed that those who had the most positive attitudes towards the selection system performed slightly better on tests, perhaps not surprisingly.

4.12 SUMMARY

This chapter has dealt with important principles in personnel selection and how selection methods, including tests, should be evaluated. Development of a selection system should start with a job analysis where the abilities and personal characteristics needed to accomplish the job should be specified. Then, methods suited to precise measurement of those capabilities or characteristics should be chosen. It is important to have evidence that the methods have predictive validity, either from local validation studies or meta-analyses. Other important aspects of the selection process are the applicants' reactions and attitudes and that the methods used are fair. The utility of the selection can be calculated in dollars or in terms of correct decisions. Often, the costs associated with conducting a selection process are small relative to the costs of employees who are not performing well or who are unable to complete training.

RECOMMENDED READING

Cook, M. 1998. *Personnel selection,* 3rd ed. Chichester, England: John Wiley & Sons.
Hunter, D. R., and Burke, E. F. 1995. *Handbook of pilot selection.* Aldershot, England: Avebury Aviation.
Mabon, H. 2004. *Arbetspsykologisk testing* [*Employment testing*]. Stockholm: Psykologi-förlaget AB.

REFERENCES

Bartram, D. 1995. Validation of the MICROPAT battery. *International Journal of Selection and Assessment* 3:83–94.
———. 2005. The great eight competences: A criterion-centric approach to validation. *Journal of Applied Psychology* 90:1185–1203.
Bartram, D., and Baxter, P. 1996. Validation of the Cathay Pacific Airways pilot selection program. *International Journal of Aviation Psychology* 6:149–169.
Bartram, D., and Dale, H. C. 1982. The Eysenck personality inventory as a selection test for military pilots. *Journal of Occupational Psychology* 55:287–296.
Brehmer, B. 2003. Predictive validation of the MRU battery. *Proceedings of the Second EUROCONTROL selection seminar.* (HRS/MSP-002-REP-07). Brüssel: EUROCONTROL.
Broach, D., and Manning, C. A. 1997. Review of air traffic controller selection: An international perspective (DOT/FAA/AM-97/15). Washington, D.C.: Federal Aviation Administration, Office of Aviation Medicine.
Burke, E., Kokorian, A., Lescreve, F., Martin, C. J., Van Raay, P., and Weber, W. 1995. Computer-based assessment: A NATO survey. *International Journal of Selection and Assessment* 3:75–83.
Callister, J. D., King, R. E., and Retzlaff, P. D. 1999. Revised NEO personality inventory profiles of male and female U.S. Air Force pilots. *Military Medicine* 164:885–890.
Carretta, T. R. 1997. Male–female performance on U.S. Air Force pilot selection tests. *Aviation, Space and Environmental Medicine* 68:818–823.
Carretta, T. R., and Ree, M. J. 2003. Pilot selection methods. In *Principles and practice of aviation psychology,* ed. P. S. Tsang and M. A. Vidulich, 357–396. Mahwah, NJ: Lawrence Erlbaum Associates.
Carretta, T. R., Rodgers, M. N., and Hansen, I. 1996. The identification of ability requirements and selection instruments for fighter pilot training (technical report no. 2). Euro–Nato Aircrew Human Factor Working Group.
Cook, M. 1998. *Personnel selection,* 3rd ed. Chichester, England: John Wiley & Sons.
Dierdorff, E. C., and Wilson, M. A. 2003. A meta-analysis of job analysis reliability. *Journal of Applied Psychology* 88:635–646.
Dockeray, F. C., and Isaacs, S. 1921. Psychological research in aviation in Italy, France, England, and the American Expeditionary Forces. *Journal of Comparative Psychology* 1:115–148.

Edgar, E. 2002. Cognitive predictors in ATCO selection: Current and future perspectives. In *Staffing the ATM system. The selection of air traffic controllers,* ed. H. Eißfeldt, M. C. Heil, and D. Broach, 73–83. Aldershot, England: Ashgate.

Eißfeldt, H. 1991. DLR selection of air traffic control applicants. In *Human resource management in aviation,* ed. E. Farmer, 37–49. Aldershot, England: Avebury Technical.

———. 1998. The selection of air traffic controllers. In *Aviation psychology: A science and a profession,* ed. K. M. Goethers, 73–80. Aldershot, England: Ashgate.

Eißfeldt, H., and Heintz, A. 2002. Ability requirements for DFS controllers: Current and future. In *Staffing the ATM system. The selection of air traffic controllers,* ed. H. Eißfeldt, M. C. Heil, and D. Broach, 13–24. Aldershot, England: Ashgate Publishing Limited.

Embretson, S. E., and Reise, S. P. 2000. *Item response theory for psychologists.* Mahwah; NJ: Lawrence Erlbaum Associates.

Fitts, P. M. 1946. German applied psychology during World War 2. *American Psychologist* 1:151–161.

Flanagan, J. C. 1954. The critical incident technique. *Psychological Bulletin* 51:327–358.

Fleishman, E. A. 1975. Toward a taxonomy of human performance. *American Psychologist* 30:1127–1149.

Goeters, K. M., and Maschke, P. 2002. Cost-benefit analysis: Is the psychological selection of pilots worth the money? Bidrag presentert på the *25th Conference of EAAP,* Warsaw, Poland, September 16–20.

Goeters, K. M., Maschke, P., and Eißfeldt, H. 2004. Ability requirements in core aviation professions: Job analysis of airline pilots and air traffic controllers. In *Aviation psychology: Practice and research,* ed. K. M. Goeters, 99–119. Aldershot: Ashgate Publishing Limited.

Hätting, H. J. 1991. Selection of air traffic control cadets. In *Handbook of military psychology,* ed. R. A. Galand and A. D. Mangelsdorff, 115–148. Chichester, England: John Wiley & Sons.

Henmon, V. A. C. 1919. Air service tests of aptitude for flying. *Journal of Applied Psychology* 2:103–109.

Hilton, T. F., and Dolgin, D. L. 1991. Pilot selection in the military of the free world. In *Handbook of military psychology,* ed. R. Gal and A. D. Mangelsdorff, 81–101. New York: John Wiley & Sons.

Hörmann, H., and Maschke, P. 1996. On the relation between personality and job performance of airline pilots. *International Journal of Aviation Psychology* 6:171–178.

Hunter, D. R. 1989. Aviator selection. In *Military personnel measurement: Testing, assignment, evaluation,* ed. M. F. Wiskoff and G. F. Rampton, 129–167. New York: Praeger.

Hunter, D. R., and Burke, E. F. 1987. Computer-based selection testing in the Royal Air Force. *Behavior Research Methods, Instruments, & Computers* 19:243–245.

———. 1994. Predicting aircraft pilot-training success: A meta-analysis of published research. *International Journal of Aviation Psychology* 4:297–313.

———. 1995. *Handbook of pilot selection.* Aldershot, England: Avebury Aviation.

Hunter, J. E., and Schmidt, F. L. 2003. *Methods of meta-analysis: Correcting error and bias in research findings.* Beverly Hills, CA: Sage.

Hunter, J. E., Schmidt, F. L., and Hunter, R. 1979. Differential validity of employment tests by race: A comprehensive review and analysis. *Psychological Bulletin* 86:721–735.

Hunter, M. 2004. Kapittel 11, Personalekonomiska aspekter. In *Arbetspsykologisk testning,* 2nd ed., ed. M. Hunter, 355–386. Stockholm: Psykologiforlaget AB.

Jessup, G., and Jessup, H. 1971. Validity of the Eysenck personality inventory in pilot selection. *Occupational Psychology* 45:111–123.

Johnson, J. T., and Ree, M. J. 1994. Rangej: A Pascal program to compute the multivariate correction for range restriction. *Educational and Psychological Measurement* 54:693–695.

Kantor, J. E., and Carretta, T. R. 1988. Aircrew selection systems. *Aviation, Space, and Environmental Medicine* 59:A32–A38.

Kelly, G. 1955. *The psychology of personal constructs*. New York: Norton.

King, R. E., Retzlaff, P. D., and McGlohn, S. E. 1998. Female United States Air Force pilot personality: The new right stuff. *Military Medicine* 162:695–697.

Kragh, U. 1960. The defense mechanism test: A new method for diagnosis and personnel selection. *Journal of Applied Psychology* 44:303–309.

Lambirth, T. T., Dolgin, D. L., Rentmeister-Bryant, H. K., and Moore, J. L. 2003. Selected personality characteristics of student naval aviators and student naval flight officers. *International Journal of Aviation Psychology* 13:415–427.

Lang-Ree, O. C., and Martinussen, M. 2006. Applicant reactions and attitudes towards the selection procedure in the Norwegian Air Force. *Human Factors and Aerospace Safety* 6:345–358.

Lawley, D. N. 1943. A note on Karl Pearson's selection formulae. *Proceedings of the Royal Society of Edinburgh* 62 (section A, part 1):28–30.

Martinussen, M. 1996. Psychological measures as predictors of pilot performance: A meta-analysis. *International Journal of Aviation Psychology* 1:1–20.

Martinussen, M., Jenssen, M., and Joner, A. 2000. Selection of air traffic controllers: Some preliminary findings from a meta-analysis of validation studies. *Proceedings of the 24th EAAP (European Association for Aviation Psychology) Conference*.

Martinussen, M., and Torjussen, T. 1993. Does DMT (defense mechanism test) predict pilot performance only in Scandinavia? In *Proceedings of the Seventh International Symposium on Aviation Psychology*, ed. R. S. Jensen and D. Neumeister, 398–403. Columbus: Ohio State University.

———. 1998. Pilot selection in the Norwegian Air Force: A validation and meta-analysis of the test battery. *International Journal of Aviation Psychology* 8:33–45.

———. 2004. Initial validation of a computer-based assessment battery for pilot selection in the Norwegian Air Force. *Human Factors and Aerospace Safety* 4:233–244.

Maschke, P. 2004. The acceptance of ab initio pilot selection methods. *Human Factors and Aerospace Safety* 4:225–232.

Melton, R. S. 1954. Studies in the evaluation of the personality characteristics of successful naval aviators. *Journal of Aviation Medicine* 25:600–604.

Moser, U. 1981. Eine Methode zure Bestimmung Wiederstandsfähigkeit gegenüber der Konfliktreaktivierung unter Verwendung des Rorschachtests, dargestellt am Problem der Pilotenselektion. *Schweizerische Zeitschrift für Psychologie* 40:279–313.

Nye, L. G., and Collins, W. E. 1993. Some personality and aptitude characteristics of air traffic control specialist trainees. *Aviation Space and Environmental Medicine* 64:711–716.

Ones, D. S., Viswesvaran, C., and Reiss, A. D. 1996. Role of social desirability in personality testing for personnel selection: The red herring. *Journal of Applied Psychology* 81:660–679.

Riis, E. 1986. Militærpsykologien i Norge. *Tidsskrift for Norsk Psykologforening* 23(Suppl. 1):21–37.

Rothstein, H. R., and McDaniel, M. A. 1992. Differential validity by sex in employment settings. *Journal of Business and Psychology* 7:45–62.

Schmidt, F. L., and Hunter, J. E. 1998. The validity and utility of selection methods in personnel psychology: Practical and theoretical implications of 85 years of research findings. *Psychological Bulletin* 124:262–274.

Schroeder, D. J., Broach, D., and Young, W. C. 1993. Contribution of personality to the prediction of success in initial air traffic control specialist training (DOT/FAA/AM-93/4). Washington, D.C.: Federal Aviation Administration, Office of Aviation Medicine.

Sells, S. B. 1955. Development of a personality test battery for psychiatric screening of flying personnel. *Journal of Aviation Medicine* 26:35–45.

———. 1956. Further developments on adaptability screening for flying personnel. *Aviation Medicine* 27:440–451.

Siem, F. M., and Murray, M. W. 1994. Personality factors affecting pilot combat performance: A preliminary investigation. *Aviation, Space and Environmental Medicine* 65:A45–A48.

Taylor, H. C., and Russel, J. T. 1939. The relationship of validity coefficients to the practical effectiveness of tests in selection. *Journal of Applied Psychology* 32:565–578.

Termøhlen, J. 1986. Flyvepsykologiens udvikling. In *Udviklingslinier i dansk psykologi: Fra Alfred Lehmann til i dag,* ed. I. K. Moustgaard and A. F. Petersen, 169–181. København: Gyldendal.

Thorndike, R. L. (1949). *Personnel selection: Test and measurement techniques.* New York: John Wiley & Sons.

Torjussen, T. M., and Hansen, I. 1999. The Norwegian defense, best in test? The use of aptitude tests in the defense with emphasis on pilot selection. *Tidsskrift for norsk psykologforening* 36:772–779.

Torjussen, T. M., and Værnes, R. 1991. The use of the defense mechanism test (DMT) in Norway for selection and stress research. In *Quantification of human defense mechanisms,* ed. M. Olff, G. Godaert, and H. Ursin, 172–206. Berlin: Springer–Verlag.

Whitener, E. M. 1990. Confusion of confidence intervals and credibility intervals in meta-analysis. *Journal of Applied Psychology* 75:315–321.

5 Training

5.1 INTRODUCTION

In the earliest days of aviation there were no instructor pilots. The first aviators, such as Orville and Wilbur Wright, Octave Chanute, Otto Lilienthal, and the Norwegian Hans Fleischer Dons* (originally a submarine officer), trained themselves. They were simultaneously test pilots and student pilots, with the inevitable consequence that many died during the process of discovering how to maintain control of their aircraft (including Lilienthal, who died in a glider crash in 1896). Aspiring modern pilots are fortunate to be the beneficiaries of the experiences of these pioneers, along with several succeeding generations of pilots who have also made their contributions to the art and science of aviation training. However, not all advances in aviation training have come from pilots. Researchers (most of them not pilots) in the fields of psychology and education have also helped shape the format, if not the content, of current aviation training. Principles of how humans learn new skills developed in the laboratories have been applied advantageously to pilot training. In this chapter, we will examine some of those principles and how they are applied in an aviation setting, along with the general process of training development.

Training is a broad term that covers a number of activities conducted in a variety of settings. Training can be categorized according to when it occurs—for example, initial training required to impart some new skill set as opposed to remedial training required to maintain those skills. It can also be categorized according to where it occurs—for example, in the classroom, in the simulator, or in an aircraft. Training can also be categorized according to the content—for example, whether it addresses purely technical issues, such as the computation of weight and balance, as opposed to nontechnical issues, such as crew coordination. Regardless of how one chooses to categorize training, the goals of these activities are common: the development of a set of skills and knowledge in the trainee to some specified level of competency.

> Training: The systematic process of developing knowledge, skills, and attitudes; activities leading to skilled behavior.

The processes and methods used to achieve the desired training goals are informed and shaped by the scientific research dealing with human learning. Beginning with the late nineteenth century work of Ebbinghaus, a great deal of research has been

---

* For more information on this and other early Norwegian aviators, consult the very interesting book, *100 Years of Norwegian Aviation* (National Norwegian Aviation Museum 2005).

directed at understanding the conditions that influence human learning. This is an immense body of work far beyond the scope of this chapter. A large number of introductory volumes on human learning are available for the student who desires more information. Recent examples include Hergenhahn and Olson (2005), Ormrod (2007), and Mazur (2006). In addition to these general works, there are works dedicated primarily to the issues of training in aviation. Some examples here include Henley (2004), O'Neil and Andrews (2000), and Telfer and Moore (1997).

Later in this chapter, we will examine some of the training issues specific to aviation. However, it is important to note that training, like the design of the physical apparatus described in an earlier chapter, should be thought of as a system. The design of effective training, particularly when such training will be used by multiple students and instructors over an extended period, requires careful consideration of a number of interrelated factors. Therefore, before we consider the specifics of aviation training, we should examine these factors and the techniques that have been developed to ensure successful training system development.

5.2 TRAINING SYSTEM DESIGN

The rigor that is applied to the design of a training system reflects the planned application of the training system. That is, very little rigor is typically placed on design of training that will be used with only one or two people for a job of little significance. Most people have experienced training of this sort. For example, a new employee is shown, in a training program lasting about 1 minute, how to operate the office copier. There is no formal training plan. There is no formal evaluation of student retention of the material. Furthermore, there is no assurance that this employee will receive the same instruction as the next employee. Standardization is lacking because standardization and the planning that it entails are expensive, and the result of failure to perform the task successfully is simply a few sheets of wasted paper.

Contrast this with the consequences of failure to train a pilot properly to execute an instrument approach or to execute a rejected takeoff. Here the consequences are potentially severe, both in terms of money and in terms of human life. Clearly, the latter situation requires a more formal approach to the design of a training program to ensure that all the necessary elements are addressed and that the student achieves a satisfactory level of performance. The expense of a rigorously developed training program is justified by the consequence of failure and, for many organizations, such as the military services, by the large numbers of personnel to be trained over an extended period.

One method for ensuring that rigor is applied is the use of the systems approach to training (SAT), also known as instructional system development (ISD).* This set of procedures to be used for the development of training systems originated with the American military in the mid-1970s. It is described in detail in a multiple volume

---

* Much of this discussion of SAT/ISD is derived from Air Force Manual 36-2234 (U.S. Air Force 1993), to which the interested reader is directed for more detailed information. Meister (1985) is also a good source of general information.

military handbook (U.S. Department of Defense 2001) and in manuals produced by each of the individual services. For example, a manual produced by the U.S. Air Force (1993) adapts the general guidance given in the Department of Defense handbook to the specific needs of the Air Force. Similarly, individualized documents exist for the U.S. Navy (NAVEDTRA 130A; 1997) and the U.S. Army (TRADOC Pamphlet 350-70; n.d.).

This structured approach to the development of training has also been embraced by civil aviation. For example, its principles are included in the Advanced Qualification Program (AQP) promoted by the Federal Aviation Administration, and in the Integrated Pilot Training program established by Transport Canada. These advances reflect the general approach of embedding the systems approach to training that is reflected in the International Civil Aviation Organization Convention on International Civil Aviation—Personnel Licensing (ICAO n.d.)

To put it simply, SAT/ISD is a process that provides a means to determine

- who will be trained;
- what training will involve;
- when training will take place;
- where training will take place;
- why training is being undertaken; and
- how training is accomplished.

The first reaction of many pilots and nonpilots when seeing such a list is

Why all the fuss? Why is an elaborate system needed for such simple questions? Surely, the answers to these questions are obvious:

- Whoever wants to be a pilot will get training.
- We will show trainees how to fly the airplane.
- We will fly in the morning or afternoon.
- The training will take place in the aircraft.
- We are doing the training so that the student can fly the plane.
- The instructor will show how it is done, and then trainees will do it themselves.

The other reaction is "Why not just keep on doing what we have always done? After all, we have been training pilots for years."

These concerns are understandable and may even have merit in some situations; however, they reflect a generally limited view of the world. If one is considering only a single instructor and a few aspiring student pilots, then the training will take place more or less as outlined in the second set of bullets. It is a casual approach to a situation that does not demand high efficiency of training or rigorous quality control of the product.

Consider, however, the situation faced by almost every military service and by many air carriers. The military take large numbers of recruits with no previous military experience and, in most cases, who lack the technical skills and knowledge required to perform the duties of their military specialty. In a relatively short time, these recruits must receive military and technical training that will enable them to function as part of a military unit, where the consequences of failure to perform are often very high. Clearly, their training cannot be left to chance. If for no other reason than to minimize the enormous costs associated with large-scale military training (e.g., the cost of producing one helicopter pilot in the U.S. Army is approximately one million U.S. dollars; Czarnecki 2004), the training programs must be designed so as to deliver exactly what is needed in a format and at a time that ensure the trainees achieve a satisfactory level of competence. To achieve this goal requires careful analysis and planning, and these are precisely what SAT/ISD provides.

The five stages of the SAT/ISD process are

- analyze;
- design;
- develop;
- implement; and
- evaluate.

> Task: "A single unit of specific work behavior, with clear beginning and ending points, that is directly observable or otherwise measurable. A task is performed for its own sake, that is, it is not dependent upon other tasks, although it may fall in a sequence with other tasks in a mission, duty, or job" (U.S. Department of Defense, 2001, p. 47).

5.2.1 Analyze

The SAT/ISD process begins with an analysis of the job for which a person is to be trained. The objective is to develop a complete understanding of what it takes to perform the job. During this stage, a task inventory is compiled in which all the tasks associated with the job are listed. In addition, the standards, conditions, performance measures, and any other criteria that are needed to perform each of the tasks on the task inventory must also be identified.

This job analysis is typically performed by observing personnel (job incumbents) on the job and making note of what they do or by interviewing incumbents about what they do. In the military services in particular, more formal occupational survey and analysis procedures may be used. One example of an instrument used to capture such information is the critical incident technique (CIT; Flanagan 1954). Once a list of the tasks performed as part of a job is compiled, the CIT procedures can be used to identify tasks that are preeminent in terms of their frequency, difficulty, and failure consequence. Clearly, more attention should be devoted in the training process to tasks that are difficult, that occur frequently, and that have dire consequence if not performed properly than to tasks that are seldom completed, are easy, and have little impact. However, until this comprehensive list of the tasks that comprise a job has been compiled and each task analyzed, there is no basis for deciding which tasks are important tasks and which are insignificant.

In addition to examining the job and its constituent tasks, attention must also be paid to the eventual recipients of the training. This is referred to as the target audience: the people who will complete the training and then go on to do the job. Just as the design of the hardware must consider the capabilities and limitations of the users, so must the training system be designed with the eventual users in mind. During the analysis phase, these users must be identified and described in detail, and this information will be critical in the design phase. Consider, for example, the importance of knowing whether the training system is to be designed to accommodate student pilots with no prior flying experience or whether it will be used by experienced commercial pilots who already have mastered the basics. This is an extreme example, perhaps, but the point is that designers of training systems cannot make assumptions about the characteristics of those who will use the training.

As a further example, in the international world of aviation, a minimum command of the English language is required. Training designers would be ill advised to assume that all the students participating in a new training course have a satisfactory command of English or even that they all have the same level of fluency. Clearly, there are substantial national and regional differences in the teaching of English. A thorough analysis would therefore assess this issue to determine whether remedial instruction in English is required for a specified target audience before the technical training may commence. Similar comments could be made with regard to computer literacy, experience in driving automobiles, prior mechanical experience, and intellectual level—to name but a few of the many possible examples. All of these predecessor, or enabling, conditions must be identified during the analysis phase.

5.2.2 DESIGN

In this phase, the instructional strategies are determined and the instructional methods and media are selected. One instructional strategy might be not to provide formal training at all. Rather, one might elect to use some sort of apprenticeship and have all learning take place as on-the-job training. This is a rather common strategy, particularly for jobs requiring lower skill levels. Consider, for example, the training program for a carpenter's helper. This might be as simple as telling the new helper to follow the carpenter and do what he or she says to do. Admittedly, this is a rather extreme example and not likely to be found in aviation settings. However, a strategy that combines some formal training with on-the-job training is fairly common. This is particularly true for tasks that are seldom performed.

The occupational analysis procedures used in the analysis phase may result in the identification of some tasks that are rarely performed. These may even be tasks with significant consequences, but which occur so infrequently that any training provided during initial qualification would be lost by the time it became necessary to perform the task. (The issue of skill decay will be discussed in some detail later in this chapter.) The instructional system designers are thus placed in a quandary. Do they include instruction on the performance of a task that they know will be forgotten, on average, long before the task must be performed? Do they rely upon refresher training to maintain task proficiency? Do they utilize some sort of just-in-time training scheme, so that when the need for the task arises, the incumbent can quickly learn the required skills?

In some instances, the latter approach is satisfactory, particularly when dealing with maintenance of highly reliable electronic systems. The reader may reflect upon the last (if any) time he or she was called upon to install or replace the disk drive in a personal computer. Without the benefit of any technical training, computer owners are called upon to accomplish this task by the manufacturers of computer disk drives. They are able to do so (usually without injury to themselves or their personal computer) due in large part to the well-designed, step-by-step instructions, with accompanying graphics, provided by disk manufacturers. The instructions provide just-in-time training to enable the personal computer owner to accomplish the required task. Thus, a person who has never accomplished this task before and has received no training can, during the course of the hour or so required to read and follow the directions, successfully complete the installation and then promptly forget everything just learned until the next occasion arises. At that time, the abbreviated training, task performance, and skill decay cycle will be repeated.

Of course, in some instances, the task demands require that even rarely performed tasks be trained to a high level of performance and that frequent refresher training be given to maintain skill levels. In aviation, the rejected takeoff (RTO) is a prime example of such a task. In such a time-critical situation, task performance must be immediate and flawless. During the 2 seconds or so in which the RTO decision must be reached and an appropriate action initiated, there is no time available to consult even the ubiquitous checklist, let alone pull out the abnormal procedures handbook and learn how to perform the task.

During the SAT/ISD analysis phase, these tasks and their performance constraints should have been identified so that appropriate instructional strategies may be selected. In each case, the strategy should be appropriate for the task for which the student is being trained.

In addition to the instructional strategy, the instructional developer must also select the instructional methods and media during this phase. Some of the instructional methods include:

- lecture;
- demonstration;
- self-study;
- computer-based (CB) training; and
- on-the-job training (OJT).

The media available include:

- printed media;
- overhead transparencies;
- audio tape recordings;
- 35 mm slide series;
- multimedia presentations;

- video and film; and
- interactive courseware.

The selection of the instructional method and media forms part of the plan of instruction developed during this phase. This plan is focused on learning objectives. These are statements of what is expected to be accomplished at each stage of the training program, and they proceed in such a sequence that at each stage the skills and knowledge required for successful completion of the new learning objective have already been put into place by the previous learning activities. Therefore, if the objective of a certain set of instructions is for the student to complete an instrument approach successfully, then at some earlier stage the student must have learned how to tune the radio, how to maintain aircraft attitude by reference to the attitude indicator, and how to initiate and control a descent, along with many other skills. Thus, the order in which new material is presented is critical to success and must be considered carefully in the plan of instruction. The analysis phase will have resulted in the identification of these predecessor skills and knowledge, and the instructional designer must ensure that these restrictions are considered during the design phase.

5.2.3 DEVELOP

Once the objectives have been established and the training strategies and activities have been planned, it is time to implement the design by creating a formal course syllabus, writing lessons, producing the instructional materials, and, if necessary, developing interactive courseware. The key document to be created is the plan of instruction (POI) or course syllabus. It is this document that serves to control the planning, organization, and conduct of the instruction. It is the blueprint for providing instruction in a course and is used to develop the individual lesson plans used by instructors in the delivery of instruction.

Perhaps the most visible, or at least the most voluminous, product of this phase is the actual instructional material. During this phase, the books, pamphlets, student guides, videotapes, slides, transparencies, simulators, mock-ups, and everything else identified during the design phase and listed in the POI are produced.

Also constructed during this phase are the tests that will determine whether the students have achieved the mastery levels specified in the POI. This is a very important component and the development of these tests must adhere to sound psychometric principles. Because these issues will be covered at some length in Chapter 2 and Chapter 4 of this book, they will not be addressed here, other than to note that the care given to tests used to select personnel for pilot training must also be applied to the tests that determine whether, at the end of training, they are now qualified to be pilots.

As one final activity in this phase, it is always wise to try out the new training program on a limited basis to ensure that everything proceeds according to plan. It is said that no battle plan survives the first contact with the enemy. Similarly, even in the best planned training program, it is almost inevitable that some things will have been overlooked; this will become glaringly obvious when the full course is administered to real students. Implicit assumptions may have been made about student capabilities that turn out to be faulty. Estimates of the time or number of trials required to learn some new skill to criterion levels may have been too optimistic. Language that seemed perfectly clear to the developers and subject matter experts may prove hopelessly confusing to naive students.

Even though each of the individual sections and training components may have been tried out with students as they were being developed, one final test of the entire system is prudent. It is here that the effects of dependencies among the training elements may be revealed that were not evident when each of the elements was evaluated in isolation. Success in this trial provides the trigger to move to the next phase with confidence.

5.2.4 IMPLEMENT

It is at this stage that the new training program becomes operational. If all the preceding stages have been accomplished successfully, then the expected students will arrive at the proper locations. The instructional materials will be on hand in sufficient quantities, and all hardware, such as simulators, will be operational. The instructors and support personnel will be in place and ready to begin instruction. This marks the boundary, then, between the largely technical activities of the preceding phase and what is now mainly a management or administrative activity.

5.2.5 EVALUATE

When conducted properly, evaluation is ongoing during the SAT/ISD process. Each of the preceding phases should include some sort of evaluation component. For example, during the analysis phase, some method is needed to ensure that all the tasks comprising the job have been included in the analysis and that appropriate, rigorous, task analytic methods have been applied to ensure a quality result. During the design phase, the selection of methods and media should be evaluated against the learning objectives to ensure that they are appropriate. During the development phase, the instructional material being created must be checked carefully for validity. For example, the radio phraseology that is taught during flight training must be checked against the phraseology prescribed by the civil or military authority because departures from standard phraseology lead to confusion and errors. Finally, at the implementation phase, several evaluation components, such as the quality of the graduates, are possible. Overall, these evaluation activities can be divided into three general types:

- formative evaluation;
- summative evaluation; and
- operational evaluation.

The formative evaluation process extends from the initial SAT/ISD planning through the small-group tryout. The purpose is to check on the design of the individual components comprising the instructional system. It answers the question, "Have we done what we planned to do?" That is, if the plan for achieving a specified learning objective called for the use of a partial simulation in which the student would be taught to operate the flight management system (FMS) to a specified level of competence, did that actually occur? Does the instructional system include the use of the partial simulation to achieve that particular learning objective? Do students who complete this particular training component demonstrate the level of competence required?

When this sort of evaluation is conducted early on, it allows the training developers to improve the training program while the system is still being developed, and changes can be made for the least cost. For example, discovery that students completing the FMS training cannot perform all the tasks to a satisfactory level could lead to changes in the training design to modify the training content or to provide additional time for practice on the simulator. However, if this deficiency were not discovered until late in the training development process, then the relatively simple and inexpensive changes might no longer be possible. For example, the simulator might have been scheduled for other training components, so additional practice on the existing simulators is no longer a possibility. This means that either major change must be made to the training program or additional simulation assets must be purchased. Both of these alternatives have negative cost implications.

The summative evaluation involves trying out the instructional program in an operational setting on students from the target population. The basic question answered by this evaluation is "Does the system work?" That is, does the instructional system work under operational conditions? Typically, the summative evaluation examines the training program during the operational tryout of two or three classes. This provides enough data to identify such issues as lack of adequate resources, changes needed to the schedules, inadequacies of the support equipment, need for additional training of instructors, or modifications to the training materials to improve clarity. The summative evaluation also addresses the key question of graduate performance. That is, it should show whether the graduates of the training program can perform the jobs for which they are being trained. Clearly, just as the graduates of a medical college should be able to pass the medical licensing examination, the graduates of a pilot training school should be able to pass the licensing authority's written and practical tests. If a significant number of graduates cannot meet these standards, then there is almost certainly something wrong in the training school.

Operational evaluation is a continuous process that should be put into place when the new training system is placed into operation. This form of evaluation is used to gather and analyze internal and external feedback data to allow management to monitor the status of the program. At a minimum, it should provide for monitoring of the status of graduates of the training course to ensure that they continue to meet job performance requirements. Changes in the proportion of graduates who pass licensing examinations, for example, should signal the need to reevaluate the training program. Perhaps some changes have taken place in the operational environment that dictate changes in the training program.

In an aviation context, the use of global positioning system (GPS) navigation provides such an example. As GPS begins to supplement or eventually even replace very high frequency omnidirectional radio (VOR) navigation, pilot training programs must be modified to provide instructions on GPS navigation techniques—perhaps at the cost of decreased instruction on VOR navigation. An alert training institution might make these changes proactively. However, the use of an ongoing operational evaluation component, particularly one that examined more than gross pass/fail rates, should signal the need for change, even if the changing technology were not otherwise noticed.

5.2.6 Conclusion

The use of SAT/ISD provides a framework for the development of training. It does not prescribe methods or modes—a feature for which it is sometimes criticized, but which others argue is a strength of the technique. Its use is ubiquitous in the military, including military aviation. Hence, some familiarity with its basic concepts is desirable, even for those who may not be directly involved in the development of a large-scale training program. An adherence to its general precepts and procedures will inform and guide any training development, even at a modest level. The casual flight instructor may not utilize a formal SAT/ISD process in planning how he or she will teach student pilots; but, at least some consideration of learning objectives, choice of teaching modes, evaluation, and the other SAT/ISD components would almost certainly result in an improved training delivery and a better trained student.

Cockpit/crew resource management (CRM): The effective use of all available resources—people, weapon systems, facilities, and equipment, and environment—by individuals or crews to safely and efficiently accomplish an assigned mission or task (definition from U.S. Air Force 2001).

5.3 CREW RESOURCE MANAGEMENT

The preceding section has dealt in some length with a generalized system for developing training, without regard for the specific elements to be trained. Typically, training is thought of in conjunction with the technical skills of operating an aircraft. These include such skills as reading a map, reading and interpreting weather forecasts, accurately calculating weight and balance, and proper movement of the controls so as to accomplish a desired maneuver. However, other skills are also valuable. These skills are typically referred to under the general heading of cockpit resource management (CRM), although in Europe they may also be termed non-technical skills (NOTECHS). This area includes such things as getting along with crew members, knowing when and how to assert one's self effectively in critical situations, and maintaining situational awareness. For the most part, training in CRM presupposes competency in all the technical skills required to operate an aircraft. However, a well-developed training program for ab initio pilots, particularly one that has been developed in accordance with the precepts of SAT/ISD, may well include CRM as a stand-alone element or as part of the technical skill training.

Ab initio: Latin for "from the beginning." This refers to individuals with no previous experience or training in the relevant subject matter.

Situation awareness: "Situation awareness is the perception of the elements in the environment within a volume of time and space, the comprehension of their meaning and the projection of their status in the near future" (Endsley 1988, p. 97).

The need for CRM training was identified in the 1980s as a result of studies of the causes of predominantly civilian airliner accidents. Christian and Morgan (1987) reviewed the causes of aircraft accidents and found that the following human factors contributed to mishaps:

- preoccupation with minor mechanical irregularities;
- inadequate leadership and monitoring;
- failure to delegate tasks and assign responsibilities;
- failure to set priorities;
- failure to communicate intent and plans;
- failure to utilize available data; and
- failure to monitor other crew members in the cockpit adequately.

Perhaps the seminal article on this subject, however, is that of Foushee (1984), in which he applied the techniques and terminology of social psychology to the environment of an air carrier cockpit. From that point, interest in CRM has mushroomed. The evolution of CRM in commercial aviation is documented by Helmreich, Merritt, and Wilhelm (1999). However, a vast literature on this subject now extends beyond aviation to such settings as the hospital operating theater (cf. Fletcher et al. 2003) and the control rooms of off-shore drilling platforms (cf. Salas, Bowers, and Edens 2001).

A major proponent of CRM has been the Federal Aviation Administration (FAA) in the United States. The FAA has sponsored an extensive program of research on CRM, most notably by Robert Helmreich and his colleagues at the University of Texas. Based on the results of the research of Helmreich and many others, the FAA has produced publications that provide guidance and definitions to air carriers and others regarding the desired characteristics of CRM training. According to the CRM advisory circular produced by the FAA (2004, p. 2):

CRM training is one way of addressing the challenge of optimizing the human/machine interface and accompanying interpersonal activities. These activities include team building and maintenance, information transfer, problem solving, decision-making, maintaining situation awareness, and dealing with automated systems. CRM training is comprised of three components: initial indoctrination/awareness, recurrent practice and feedback, and continual reinforcement.

That same FAA advisory circular provides a listing of the characteristics of effective CRM:

- CRM is a comprehensive system of applying human factors concepts to improve crew performance.
- CRM embraces all operational personnel.
- CRM can be blended into all forms of aircrew training.
- CRM concentrates on crew members' attitudes and behaviors and their impact on safety.
- CRM uses the crew as the unit of training.
- CRM is training that requires the active participation of all crew members. It provides an opportunity for individuals and crews to examine their own behavior and to make decisions on how to improve cockpit teamwork.

5.3.1 GENERATIONS OF CRM

In their review of the evolution of CRM, Helmreich et al. (1999) identified five distinct generations of CRM, beginning with the first comprehensive CRM program begun by United Airlines in 1981. These authors (p. 20) characterize first-generation courses as "psychological in nature, with a heavy focus on psychological testing and such general concepts as leadership." Perhaps not surprisingly, some pilots resisted some of these courses as attempts to manipulate their personalities.

The second generation of CRM courses is characterized as having more of a focus on specific aviation concepts related to flight operations. In addition, the training became more modular and more team oriented. This is reflected in the change of names from "cockpit resource management" in the first generation to "crew resource management." This training featured much more emphasis on team building, briefing strategies, situation awareness, and stress management. Although participant acceptance was greater for these courses, some criticism of the training and its use of psychological jargon remained. Helmreich et al. note that second-generation courses continue to be used in the United States and elsewhere.

In the third generation of CRM courses, beginning in the early 1990s, the training began to reflect more accurately and holistically the environment in which the aircrews operate. Thus, organizational factors such as organization culture and climate began to be included. This may reflect to some degree the writings by Reason (1990) at about that time, in which he described the multiple-layer concept of accident causality. In that model (to be described in more detail in Chapter 8 on safety), organizational factors are clearly identified as possible contributors to accidents.

The initiation of the advanced qualification program (AQP; Birnbach and Longridge 1993; Mangold and Neumeister 1995) by the FAA in 1990 marked the beginning of the integration and proceduralization of CRM that define the fourth generation. AQP allows air carriers to develop innovative training to meet their particular needs. However, the FAA requires that both CRM and line-oriented flight training (LOFT) be included as part of the AQP for an air carrier. Because AQP provided an air carrier advantages in terms of customizing training so as to reduce costs while maintaining a satisfactory product, most air carriers have adopted the program. Accordingly, they have also developed comprehensive CRM programs based on detailed analyses of training requirements and human factors issues.

Finally, Helmreich et al. suggest that a fifth generation of CRM training may be characterized by an explicit focus on error management. They propose that the ultimate purpose of CRM, which should be reflected in the training syllabus and the training exercises, is to develop effective means to manage risks. This philosophy reflects the Reason (1990, 1997) model of accident causality referred to earlier, in which it is acknowledged that perfect defenses against accidents do not exist. That is, despite the best laid plans and best designed systems, there will inevitably be failures, mistakes, and errors. It is prudent, therefore, to train to expect, recognize, and manage those risks.

Thus, CRM has progressed from an early emphasis on teaching good interpersonal relations to its current focus on effectively utilizing all the resources at the disposal of a flight crew (other flight-deck crew, cabin attendants, dispatchers, air traffic control, ground maintenance staff, etc.) to manage risk. In making this transition, CRM has moved from a program with sometimes only vaguely defined goals to a more sharply defined program with well-defined behavioral markers and well-established assessment procedures. Even so, the debate over whether CRM works continues. That is, does CRM actually result in improved safety?

5.3.2 EVALUATION OF CRM EFFECTIVENESS

The effectiveness of CRM has been debated in the scientific literature and on the flight deck since its inception. As noted previously, some early participants in the training resisted it because of its predominantly psychological nature, which they perceived as an attempt to manipulate their personalities. Changes in the training format and, to some extent, its content have largely eliminated these criticisms. However, even training that is well received and liked by participants may or may not achieve the desired effect.

The effectiveness of CRM has been investigated by many researchers, and it has been summarized in a study by Salas and his associates (2001) at the University of Central Florida. These researchers reviewed 58 published studies of CRM training in an aviation setting to establish its effectiveness. They used Kirkpatrick's (1976) typology for training evaluation as their framework to evaluate the effectiveness of the CRM training. The Kirkpatrick typology organizes the data obtained after training into the categories of reactions, learning, behaviors, and results (impact on organization). Of these categories, reactions to training are the easiest to collect and typically consist of the responses of participants to Likert scale statements such as "I found the training interesting." Participants indicate the strength of their agreement or disagreement with the statement by choosing one of (typically) five alternatives: strongly agree, agree, undecided, disagree, or strongly disagree. Collections of such questions can be analyzed to assess the reactions of the participants to the training for any of several dimensions (e.g., interest, relevance, effectiveness, utility, etc.).

Of the 58 studies reviewed by Salas et al., 27 involved the collection of reaction data. Their results showed that participants generally liked the CRM training and that the training that utilized role play was better liked than the lecture-based training. In addition, participants also felt that CRM training was worthwhile, useful, and applicable.

Assessments at the second level (learning) have also shown that CRM generally has a positive impact. This has been evidenced in studies of changes in both attitudes and knowledge. Attitudes toward CRM (specifically a positive attitude regarding CRM training) have been shown to be more positive following the training. In addition, increased knowledge of human factors issues, crew performance, stressors, and methods of dealing with stressors have also been observed following CRM training.

Salas et al. found that 32 studies included some assessment of behavioral change following CRM training. Most commonly, this was assessed through the measurement of CRM-related behaviors while participants engaged in a simulated flight, although in some instances (11 out of 32) an online assessment of behavior was used. Most of these studies showed that CRM training had a positive impact on behavior in that CRM-trained crews exhibited better decision making, mission analysis, adaptability, situation awareness, communication, and leadership. These findings provide strong support for the impact of CRM on crew behavior.

Although the studies of crew behavior indicate that CRM training has an impact, the effects of those changes in behavior have yet to be demonstrated clearly at an organizational level. Only six studies collected some form of evaluation data at this level, and Salas et al. (p. 651) noted that "the predominant type of evidence that has been used to illustrate CRM's impact on aviation safety consists of anecdotal reports...." Thus, a clear relationship between CRM and the desired outcome of increased safety and corresponding decrease in accidents has yet to be demonstrated. It is not surprising, then, that Salas and colleagues end their review by calling for more and better evaluations to assess the safety impact of CRM training.

That additional evaluations of CRM in terms of its method of implementation and in terms of its content are needed is made evident by continuing evidence of CRM-related contributions to accidents. In a military context, Wilson-Donnelly and Shappell (2004) reported on a study whose objective was to determine which of the CRM skills included in the U.S. Navy CRM training program matched CRM failures identified in naval aviation accidents. The following seven critical skills of CRM were included in U.S. Navy training:

- decision making;
- assertiveness;
- mission analysis;
- communication;
- leadership;
- adaptability/flexibility; and
- situational awareness.

Of the 275 Navy/Marine Corps accidents involving some CRM failure during the period 1990–2000, lack of communication was identified as the number one CRM failure, occurring in over 30% of the accidents. Inadequate briefing was the second most prevalent CRM failure, occurring in slightly over 20% of the accidents. A previous study (Wiegmann and Shappell 1999) had found that CRM failures contributed to more than half of all major (Navy Class A) accidents, so this suggests that failures associated with the materials being specifically addressed in the Navy/ Marine Corps CRM training continue to be a major factor in accidents in the Navy/ Marine Corps. Wilson-Donnelly and Shappell recognized the unsatisfactory nature of this situation, but suggested that before the current CRM training program is modified or scrapped, similar analyses should be conducted of civilian accident data to see if the same situation holds there.

Based on these analyses and investigations of the contributing factors in civilian accidents, it seems clear that not all crew resource management issues have been successfully resolved by CRM training, at least in its current implementation.

Training transfer is the "extent to which the learned behavior from the training program is used on the job" (Phillips 1991).

5.4 SIMULATOR TRAINING

Every military pilot, every airline pilot, and many instrument-rated private pilots have had some exposure to flight simulators. They are used for a number of reasons, not the least of which is the cost savings they provide over in-flight instruction. For example, the U.S. Air Force estimates that an hour of training in a C-5 aircraft costs $10,000 while an hour in a C-5 simulator is $500 (Moorman, 2002, as cited in Johnson 2005).

Particularly for the military and air carrier pilots, simulators also have the great advantage of allowing pilots to practice maneuvers and respond to events that are far too hazardous to practice in a real aircraft. The rejected takeoff decision due to loss of an engine on takeoff is an obvious example from civil aviation.

However, the utility of simulation for training is not a given. It has to be established that the training provided on a device will influence behavior in the real environment. Will what is learned in the simulator carry over into the aircraft? This is the question usually referred to as *transfer of training*, and it has been the subject of much research. Clearly, the preponderance of results shows that what pilots learn in a simulator is carried over to the aircraft.

Two major reviews of the transfer of training effectiveness in flight simulation have been conducted. The first study (Hays et al. 1992) reviewed the pilot training literature from 1957 to 1986. Using meta-analysis (see the discussion of this statistical technique in Chapter 2), Hays et al. found that simulators consistently led to improved training effectiveness for jet pilots, relative to training in the aircraft only. However, the same results were not found for helicopter pilots.

In the second major review, Carretta and Dunlap (1998) reviewed the studies conducted from 1987 to 1997. In particular, they focused on landing skills, radial bombing accuracy, and instrument and flight control. In all three of these areas, Carretta and Dunlap concluded that simulators had been shown to be useful for training pilot skills.

In one of the studies of landing skills cited by Carretta and Dunlap (1998), Lintern and colleagues (1990) examined the transfer of landing skills from a flight simulator to an aircraft in early flight training. They compared one group of pilots who were given two sessions of practice on landings in a simulator prior to the start of flight training to a control group that was given no practice prior to the start of training in the aircraft. They found that the experimental group that had received the 2 hours of simulator training required 1.5 fewer hours prior to solo than the control group. For this group, 2 hours of simulator time were equivalent to 1.5 hours of aircraft time. Comparisons of this sort are usually given in terms of the transfer effectiveness ratio (TER).

Originated by Roscoe (1980), the TER expresses the degree to which hours in the simulator replace hours in the aircraft and is defined as

$$TER = (\text{control group time} - \text{experimental group time})/\text{time of total training}$$

For example, if private pilot training normally requires 50 flight hours, and the use of a 10-hour simulator training program reduces the requirement to 40 in-flight hours, then

$$TER = (50 - 40)/10 = 1$$

In other words, 1 hour of simulator times saves 1 hour of flight time.

For a more realistic example, if the simulator training took 10 hours, and 45 additional hours of in-flight training were required, then

$$TER = (50 - 45)/10 = 0.5$$

This is interpreted to mean that 1 hour of simulation saves 0.5 hour of flight time.

Simulators are widely used for training instrument skills, and Carretta and Dunlap (1998) concluded that "simulators provide an effective means to train instrument procedures and flight control" (p. 4). They cite Pfeiffer, Horey, and Butrimas (1991), who found a correlation of $r = .98$ between simulator performance and actual flight performance.

Flight simulators can vary substantially in their fidelity to the actual flight environment in terms of motion, control dynamics, visual scene, and instrumentation. Some simulators are designed for whole-task training, whereas others are designed as part-task trainers (e.g., intended only to train students in the use of the flight management system or the aircraft pressurization system). In general, the reviews such as Carretta and Dunlap's (1998) have shown that high-fidelity simulators are not necessary for successful transfer of training.

Vaden and Hall (2005) conducted a meta-analysis to examine the true mean effect for simulator motion with respect to fixed-wing training transfer. Working with a rather small sample of only seven studies, they found a small ($d = 0.16$) positive effect for motion. (For an explanation of effect size, "d," see Chapter 2 on statistics.) Thus, although their study shows that motion does promote a greater transfer of training, the relatively small effect may not be worth the considerable expense that motion-based simulators entail, over and above a fixed-based simulator cost. Indeed, the ultimate reason for using a motion-based simulator may not be for greater transfer of training but rather for decreased simulator sickness. (For a review of simulator sickness research, see Johnson 2005).

In four quasi-experiments, Stewart, Dohme, and Nullmeyer (2002) investigated the potential of simulators to replace a portion of the primary phase of U.S. Army rotary-wing training. In their studies, positive TERs were observed for most flight maneuvers from the simulator to the UH-1 training helicopter. Generally, student pilots who received simulator training required less training to reach proficiency on flight maneuvers than controls did. For example, the TERs for the maneuver "takeoff to hover" ranged from 0.18 to 0.32 for the four experiments, while TERs for "land from hover" ranged from 0.25 to 0.72. Because the simulator was undergoing continual refinement during the course of the four experiments, Stewart et al. were able to show that improvements in the visual scene and aerodynamic flight model could result in improvements to the TER.

In a subsequent study (Stewart and Dohme 2005), the use of an automated hover trainer was investigated. This system utilized the same simulator as was used in the Stewart et al. (2002) series of experiments and incorporated a high-quality visual display system. A simple two-group design was used; 16 pilot trainees received the experimental training and 30 trainees served as controls. For the five hovering tasks that were practiced in the simulator, no instances of negative transfer of training occurred. In all cases, fewer iterations of the tasks were required for the simulator-trained subjects than for the control subjects. Stewart and Dohme conclude that these results show the potential for simulation-based training in traditionally aircraft-based primary contact tasks, in addition to its traditional role in instrument training.

In a similar experiment (Macchiarella, Arban, and Doherty 2006) using civilian student pilots at Embry-Riddle University, 20 student pilots received their initial training in a Frasca flight training device (FTD) configured to match the Cessna 172S in which 16 control subjects were trained. Positive TERs were obtained for 33 out of 34 tasks in this study. This is particularly interesting because the Frasca, in contrast to the simulators used in the Army studies, is a nonmotion-based simulator.

5.5 TRAINING USING PERSONAL COMPUTERS

In contrast to traditional simulators that generally replicate to a fair degree of realism the flight deck, instruments, and controls of a particular aircraft, several recent studies have assessed the use of personal computers for training. Although some studies of crude personal computer-based training devices (by current standards) were conducted as early as the 1970s, the major impetus for this work began in the early 1990s with the work of Taylor and his associates at the University of Illinois at Urbana-Champaign.

As with traditional simulators, the initial studies were primarily concerned with the use of personal computer aviation training devices (PCATDs) for the training of instrument skills. Taylor et al. (1999) evaluated the extent to which a PCATD could be used to teach instrument tasks and the subsequent transfer of those skills to an aircraft. They constructed a PCATD out of commercially available software and hardware and administered portions of two university-level aviation courses to students. Following instruction in the PCATD, the students received instruction and evaluation in an aircraft. TERs ranged from a high of 0.28 to a low of 0.12, depending on the specific task. An interesting finding from this study was that the PCATD was more effective for the introduction of new tasks than for the review of tasks previously learned to criterion level.

In a subsequent study, Taylor and colleagues (2001) demonstrated again that these devices can be used successfully to teach instrument skills, with an overall TER of 0.15, or a savings of about 1.5 flight hours for each 10 hours of PC-based training. An earlier study by Ortiz (1995) demonstrated a TER = 0.48 for PC-based training of instrument skills to students with no previous piloting experience.

In addition to their use in acquiring initial instrument flight skills, PCATDs might also be of use in maintaining those skills. As we note in the following section, complex cognitive skills are subject to loss over extended periods with no practice. Whether PCATDs could provide that practice was the question addressed by Talleur et al. (2003) in their study of 106 instrument-rated pilots They randomly assigned pilots to one of four groups who, following an initial instrument proficiency check (IPC) flight in an aircraft, received training at 2 and 4 months in (1) an aircraft, (2) an FTD, (3) a PCATD, or (4) none (control group). At the 6-month point, all groups then received another IPC in an aircraft.

By comparing the performance of these four groups on the final IPC, these researchers were able to demonstrate that the PCATD was effective for maintaining instrument currency. The pilots who trained on the PCATD during the 6-month period performed as well as those who trained on the FTD. Furthermore, both groups performed at least as well as those who trained in the aircraft. An additional interesting finding from this study was that, of the legally instrument-current pilots who entered this study, only 42.5% were able to pass the initial IPC in the aircraft.

Although PCATDs have been shown to be useful in the training of instrument skills, they have been less successful in dealing with manual or psychomotor skills. Two studies have failed to find transfer of manual flying skills from the PCATD to straight-and-level flight (Dennis and Harris 1998) or to aerobatic flight (Roessingh 2005). However, the use of PC-based systems to teach teamwork skills has been successfully demonstrated in a study of U.S. Navy pilots (Brannick, Prince, and Salas 2005). In that study, in a scenario executed in a high-fidelity flight simulator, pilots who received training in CRM on a PC later demonstrated better performance compared to pilots who received only problem-solving exercises and video games such as those that have been used in commercial CRM training.

5.6 RECURRENT TRAINING AND SKILL DECAY

The purpose of recurrent or refresher training is to maintain skills that have been acquired through some initial training process. Obviously, if humans never forgot anything and if motor skills could be maintained at a high level of performance indefinitely without practice, recurrent training would not be necessary. Sadly, humans forget. In fact, they forget many things rather rapidly. Even though it is said that once one learns how to ride a bicycle, one will never forget, most people will notice a decline in their bicycling proficiency after a long period of inactivity. A person may still be able to ride the bicycle, but should not try anything fancy.

Nevertheless, some skills decay faster than others. Memories and cognitive skills tend to be lost much faster than motor skills. Thus, a pilot coming back to aviation after a long absence may find that he or she can still fly the plane, but cannot remember how to get taxi and takeoff clearance and does not have a clue as to how to compute density altitude. This phenomenon is known as skill decay. Skill decay "refers to the loss or decay of trained or acquired skills (or knowledge) after periods of nonuse" (Arthur et al. 1998, p. 58).

Difficulties associated with skill decay are exacerbated by the current generation of cockpit automation that tends to place pilots in a passive, monitoring mode. However, at the time something fails, pilots must immediately take positive control of the aircraft and, in some instances, perform tasks that they have only infrequently or never been trained to accomplish. Amalberti and Wibaux (1995) point out that in many cases manual procedures are no longer taught. The example they cite is the use of the brake system during rejected takeoff on the Airbus A320. In that aircraft, use of the auto brake is mandatory. Hence, pilots are not trained to use the manual brake. It is interesting to speculate what will happen when, as all things must, the auto brake system fails, and manual braking must be used.

Prophet (1976) conducted an extensive review of the literature on the long-term retention of flying skills. His review covered some 120 sources for which abstracts or annotations were available, predominantly from military sources. His results suggest basic flight skills can be retained fairly well for extended periods of not flying. However, significant decrement occurs, particularly for instrument and procedural skills. He notes a consistent finding that continuous control (i.e., tracking) skills are retained better than the skills involved in the execution of discrete procedures. One such example is the finding by Wright (1973) that basic visual flight skills remained generally acceptable for up to 36 months, while instrument flight skills fell below acceptable levels within 12 months for about half of the pilots.

In a controlled study of the retention of flying skills (Childs, Spears, and Prophet 1983), a group of 42 employees of the Federal Aviation Administration received training necessary to qualify them for the private pilot certificate. Using a standardized assessment procedure, their proficiency was then reassessed 8, 16, and 24 months following award of their certificates. The authors reported a decline in the mean percentage of correctly performed measures, beginning with 90% and declining steadily to approximately 50% at the 24-month check. They concluded that "recently certificated private pilots who do not fly regularly can be expected to undergo a relatively rapid and significant decrement in their flight skills" (p. 41).

Childs and Spears (1986) reviewed the studies dealing with the problem of flight-skill decay. They suggested that cognitive/procedural skills are more prone than control-oriented skills to decay over periods of disuse.

Casner, Heraldez, and Jones (2006) examined pilots' retention of aeronautical knowledge in a series of four experiments in an attempt to discover characteristics of the pilots and their flying experiences that influence remembering and forgetting. They used questions from the FAA private pilot written examination.

In the first experiment, the average score for the 10-item multiple choice test was 74.8%. Of the 60 pilot participants, 12 had scores in the range of 30–60%, substantially below the minimum score (70%) required to pass the FAA certification examination. Of the 20 pilots who held a private pilot license and were not pursuing any more advanced rating, the average score was 69.5%. The national average score for the FAA private pilot written examination is 85%, so clearly some substantial forgetting of the material had occurred. The 20 certified flight instructors (CFIs), however, had an average score of 79%, suggesting that they rehearsed their knowledge more often than the other pilots did.

Although little correlation was observed between the test scores and total flight time, significant correlations were obtained between recent flight experience (previous 3 and 6 months). Most of this association may be attributed, however, to the strong correlations obtained for the CFIs ($r = .34$ and $r = .52$ for 3 and 6 months, respectively).

In their second experiment, Casner et al. asked 24 active pilots who had flown only one make and model of aircraft to perform weight and balance calculations for that aircraft and for another aircraft in which they had no prior experience. They found that whereas the pilots retained the knowledge of how to make the computations for their own aircraft, they performed considerably more poorly with the unfamiliar aircraft. Specifically, they were able to recognize a "no go" situation only 50% of the time.

Casner et al. concluded that "the certificates and ratings held by pilots have little influence on how well those pilots retain what they have learned during training" (p. 93). They suggest that there is a need for more explicit standards for ongoing aeronautical knowledge proficiency as well as some alternative methods for ensuring that pilots maintain their knowledge. In addition, they suggest that the current practices of aviation education—specifically the emphasis on abstract facts, remote from practical application—may be implicated in this failure to retain important aeronautical knowledge.

Arthur et al. (1998) used meta-analytic techniques to review the skill retention and skill decay literature. They noted that the skill decay literature has identified several factors that are associated with the decay or retention of trained skills. The most important of those factors were

- the length of the retention interval—longer intervals produce more decay;
- the degree of overlearning—overlearning aids retention and the amount of overlearning is the single most important determinant of retention; and
- task characteristics—open-loop tasks like tracking and problem solving are better retained, compared to closed-loop tasks, such as preflight checks.

From their analysis of 53 articles, Arthur and colleagues found that "physical, natural, and speed-based tasks were less susceptible to skill loss than cognitive, artificial, and accuracy-based tasks" (1998, p. 85).

These conclusions are reflected in the findings of a study on the forgetting of instrument flying skills (Mengelkoch, Adams, and Gainer 1971). In that study, two groups of 13 subjects with no prior flight experience were given academic training and instrument flying instruction in a simulator. One group received 5 training trials while the other group was given 10 training trials. Both groups were able to complete a program of maneuvers and flight procedures successfully at the conclusion of the training, although the group with the additional training performed substantially better than the other group (95% and 78% correct performance for the 10-trial and 5-trial groups, respectively). After a 4-month interval, the 10-trial group had a 16.5% loss in performance of procedures, while the 5-trial group had a 20.1% loss. Retention loss of the flight control parameters was generally much less, with only altitude and airspeed suffering a significant loss over the retention interval for both groups.

In a study by Ruffner and Bickley (1985), 79 U.S. Army aviators participated in a 6-month test period in which they flew zero, two, four, or six contact and terrain flight tasks in the UH-1 aircraft. Their results indicated that average level of performance in helicopter contact and terrain flight tasks was maintained after 6 months of no practice. Further, intervening practice flights (up to six) did not significantly improve the average level of performance. These findings were true regardless of total career flight hours or whether the tasks were psychomotor or procedural.

5.7   CONCLUDING REMARKS

Training is a vast subject that draws upon many disciplines and scientific and technical traditions, from the learning theorists to the computer scientists and simulator engineers. For the most part, our current training programs produce graduates with the skills necessary to be good pilots. Research today is largely concerned with working at the edges to improve efficiency, reduce costs, and refine training content. At the risk of doing a severe injustice to the extensive body of work that we have touched upon, sometimes only briefly, in this chapter, let us offer the following summary:

- A well-designed pilot training system requires analysis and careful planning in order to produce a quality product.
- CRM has been a topic of debate for several years, with no clear resolution in sight. Whether it actually results in improved safety is still open to question.
- Simulators are unquestionably valuable tools in aviation training and have consistently been shown to have a positive TER, in addition to saving money and allowing us to train for hazardous situations safely.
- Personal computers have made their way into aviation training. There seems to be little doubt that they can be used successfully for both initial and refresher instrument training. Whether they can also be used for contact training remains to be seen.

Finally, before we leave the subject of training, let us make a final observation about the current practice of aviation training. In particular, we are concerned with the pervasive practice of teaching subjects, such as meteorology, in a decontextualized manner. That is, the typical aviation training school offers a course on meteorology in which students learn the names of all the clouds, memorize the symbols that indicate wind speed and direction from the meteorological charts, and learn how to decipher the abbreviations contained in the METAR/TAF reports. Usually, all this takes place without any reference to the context in which such knowledge would be useful and applied. Hence, students memorize the content without learning how to apply that knowledge when planning and conducting flights. Nor do they learn to appreciate the significance of the information they are learning from an operational or safety standpoint. The need for instruction to take place within the context of its application has been discussed cogently by Lintern (1995), who refers to this concept as situated instruction.

This chapter has dealt at length with the rational processes of training system design exemplified by the SAT/ISD process; however, as Lintern (1995) notes, this process can have the effect of removing learning from the context in which it is to be applied. It is well to keep in mind that, eventually, the pilot must integrate all that he or she has learned in order to conduct a flight successfully and safely. Overcompartmentalization and a rote-learning approach to instruction, even if the instruction includes all the separate skill elements, may not provide the goal of allowing the pilot to generalize from the classroom setting to the flight deck at the time at which the knowledge must be applied. Training must always be planned and conducted with this ultimate goal in mind.

RECOMMENDED READING

Gagne, R. M., Briggs, L. J., and Wager, W. W. 1992. *Principles of instructional design.* New York: Harcourt Brace Jovanovich.
Johnston, N., Fuller, R., and McDonald, N. 1995. *Aviation psychology: Training and selection.* Aldershot, England: Avebury.
O'Neil, H. F., and Andrews, D. H. 2000. *Aircrew training and assessment.* Mahwah, NJ: Lawrence Erlbaum Associates.
Telfer, R. A., and Moore, P. J. 1997. *Aviation training: Learners, instruction and organization.* Aldershot, England: Avebury.

REFERENCES

Amalberti, R., and Wibaux, F. 1995. Maintaining manual and cognitive skills. In *Aviation psychology: Training and selection,* ed. N. Johnston, R. Fuller, and N. McDonald, 339–353. Brookfield, VT: Ashgate.
Arthur, W., Bennett, W., Stanush, P. L., and McNelly, T. L. 1998. Factors that influence skill decay and retention: A quantitative review and analysis. *Human Performance* 11:57–101.
Birnbach, R., and Longridge, T. 1993. The regulatory perspective. In *Cockpit resource management,* ed. E. Wiener, B. Kanki, and R. Helmreich, 263–282. San Diego, CA: Academic Press.
Brannick, M.T., Prince, C., and Salas, E. 2005. Can PC-based systems enhance teamwork in the cockpit? *International Journal of Aviation Psychology* 15:173–187.
Carretta, T. R., and Dunlap, R. D. 1998. Transfer of training effectiveness in flight simulation: 1986 to 1997 (technical report AFRL-HE-AZ-TR-1998-0078). Mesa, AZ: U.S. Air Force Research Laboratory, Human Effectiveness Directorate.

Casner, S. M., Heraldez, D., and Jones, K. M. 2006. Retention of aeronautical knowledge. *International Journal of Applied Aviation Studies* 6:71–97.

Childs, J. M., and Spears, W. D. 1986. Flight-skill decay and recurrent training. *Perceptual and Motor Skills* 62:235–242.

Childs, J. M., Spears, W. D., and Prophet, W. W. 1983. Private pilot flight skill retention 8, 16, and 24 months following certification (technical report DOT/FAA/CT-83/34). Washington, D.C.: Federal Aviation Administration.

Christian, D., and Morgan, A. 1987. Crew coordination concepts: Continental Airlines' CRM training. In *Cockpit resource management training,* ed. H. W. Orlady and H. C. Foushee (technical report no. NASA CP-2455). Moffett Field, CA: NASA Ames Research Center.

Czarnecki, K. R. 2004. Respect the weather. *Flightfax* 32:5–7.

Dennis, K. A., and Harris, D. 1998. Computer-based simulation as an adjunct to ab initio flight training. *International Journal of Aviation Psychology* 8:277–292.

Endsley, M. R. 1988. Design and evaluation for situation awareness enhancement. In *Proceedings of the 32nd annual meeting of the Human Factors Society,* 97–101. Santa Barbara, CA.

FAA. 2004. Crew resource management training (advisory circular 120-51E). Washington, D.C.: Federal Aviation Administration.

Flanagan, J. C. 1954. The critical incident technique. *Psychological Bulletin* 51:327–358.

Fletcher, R., Flin, R., McGeorge, P., Glavin, R., Maran, N., and Patey, R. 2003. Anesthetists' nontechnical skills (ANTS): Evaluation of a behavioral marker system. *British Journal of Anaesthesia* 90:580–588.

Foushee, H. C. 1984. Dyads and triads at 35,000 feet: Factors affecting group process and aircrew performance. *American Psychologist* 39:885–893.

Hays, R. T., Jacobs, J. W., Prince, C., and Salas, E. 1992. Flight simulator training effectiveness: A meta-analysis. *Military Psychology* 4:63–74.

Helmreich, R. L., Merritt, A. C., and Wilhelm, J. A. 1999. The evolution of crew resource management training in commercial aviation. *International Journal of Aviation Psychology* 9:19–32.

Henley, I. M. A. 2004. *Aviation education and training: Adult learning principles and teaching strategies.* Brookfield, VT: Ashgate.

Hergenhahn, B., and Olson, M. 2005. *An introduction to the theories of learning,* 7th ed. New York: Prentice Hall.

ICAO. n.d. *Convention on international civil aviation—Annex 1: Personnel licensing.* Montreal: Author.

Johnson, D. M. 2005. Introduction to and review of simulator sickness research (research report 1832). Ft. Rucker, AL: U.S. Army Research Institute.

Kirkpatrick, D. L. 1976. Evaluation of training. In *Training and development handbook: A guide to human resources development,* ed. R. L. Craig. New York: McGraw–Hill.

Lintern, G. 1995. Flight instruction: The challenge from situated cognition. *International Journal of Aviation Psychology* 5:327–350.

Lintern, G., Roscoe, S. N., Koonce, J. M., and Segal, L. D. 1990. Transfer of landing skills in beginning flight training. *Human Factors* 32:319–327.

Macchiarella, N. D., Arban, P. K., and Doherty, S. M. 2006. Transfer of training from flight training devices to flight for ab initio pilots. *International Journal of Applied Aviation Studies* 6:299–314.

Mangold, S., and Neumeister, D. 1995. CRM in the model AQP: A preview. In *Proceedings of the Eighth International Symposium on Aviation Psychology,* ed. R. S. Jensen and L.A. Rakovan, 556–561. Columbus: Ohio State University.

Mazur, J. E. 2006. *Learning and behavior,* 6th ed. New York: Prentice Hall.

Meister, D. 1985. *Behavioral analysis and measurement methods.* New York: John Wiley & Sons.

Mengelkoch, R. F., Adams, J. A., and Gainer, C. A. 1971. The forgetting of instrument flying skills. *Human Factors* 13:397–405.

Moorman, R. W. 2002. The civilian looking military. *MS&T Magazine* 6:18–20.

National Norwegian Aviation Museum. 2005. *100 years of Norwegian aviation.* Bodø, Norway: Author.

O'Neil, H. F., and Andrews, D. H. 2000. *Aircrew training and assessment.* Mahwah, NJ: Lawrence Erlbaum Associates.

Ormrod, J. E. 2007. *Human learning,* 5th ed. New York: Prentice Hall.

Ortiz, G. A., 1995. Effectiveness of PC-based flight simulation. *International Journal of Aviation Psychology* 4:285–291.

Pfeiffer, M. G., Horey, J. D., and Butrimas, S. K. 1991. Transfer of simulated instrument training to instrument and contact flight. *International Journal of Aviation Psychology* 1:219–229.

Phillips, J. 1991. *Handbook of training evaluation and measurement methods,* 2nd ed. Houston, TX: Gulf Publishing Company.

Prophet, W. W. 1976. Long-term retention of flying skills: A review of the literature (HumRRO final technical report FR-ED(P) 76-35). Alexandria, VA: Human Resources Research Organization.

Reason, J. 1990. *Human error.* New York: Cambridge University Press.

———. 1997. *Managing the risks of organizational accidents.* Aldershot, England: Ashgate.

Roessingh, J. J. M. 2005. Transfer of manual flying skills from PC-based simulation to actual flight—Comparison of in-flight measured data and instructor ratings. *International Journal of Aviation Psychology* 5:67–90.

Roscoe, S. 1980. *Aviation psychology.* Ames: Iowa State University Press.

Ruffner, J. W., and Bickley, W. R. 1985. Validation of aircrew training manual practice iteration requirements (interim report ADA173441). Fort Rucker, AL: Anacapa Science Inc.

Salas, E., Bowers, C. A., and Edens, E. 2001. *Improving teamwork in organizations: Applications of resource management training.* Mahwah, NJ: Lawrence Erlbaum Associates.

Salas, E., Burke, S., Bowers, C., and Wilson, K. 2001. Team training in the skies: Does crew resource management (CRM) training work? *Human Factors* 43:641–674.

Stewart, J. E., and Dohme, J. A. 2005. Automated hover trainer: Simulator-based intelligent flight training system. *International Journal of Applied Aviation Studies* 5:25–39.

Stewart, J. E., Dohme, J. A., and Nullmeyer, R. T. 2002. U.S. Army initial entry rotary-wing transfer of training research. *International Journal of Aviation Psychology* 12:359–375.

Talleur, D. A., Taylor, H. L., Emanuel, T. W., Rantanen, E., and Bradshaw, G. L. 2003. Personal computer aviation training devices: Their effectiveness for maintaining instrument currency. *International Journal of Aviation Psychology* 13:387–399.

Taylor, H. L., Lintern, G., Hulin, C., Talleur, D., Emanuel, T., and Phillips, A. 1999. Transfer of training effectiveness of a personal computer aviation training device. *International Journal of Aviation Psychology* 9:319–335.

Taylor, H. L., Talleur, D. A., Emanuel, T. W., Rantanen, E. M., Bradshaw, G., and Phillips, S. I. 2001. Incremental training effectiveness of personal computers used for instrument training. In *Proceedings of the 11th International Symposium on Aviation Psychology.* Columbus: Ohio State University.

Telfer, R. A., and Moore, P. J. 1997. *Aviation training: Learners, instruction and organization.* Brookfield, VT: Avebury.

U.S. Air Force. 1993. *Instructional system design* (AFMAN 36-2234). Washington, D.C.: Author.

———. 2001. Cockpit/crew resource management training program (Air Force Instruction 11-290). Washington, D.C.: Air Force Materiel Command.

U.S. Army. n.d. Systems approach to training (TRADOC pamphlet 350-70, 6 parts). Fort Monroe, VA: Author.

U.S. Department of Defense. 2001. Instructional systems development/systems approach to training and education (MIL-HDBG-29612-2A, 5 parts). Washington, D.C.: Author.

U.S. Navy. 1997. NAVEDTRA 130A. *Task based curriculum development manual (developer's guide)*, vol. 1. Pensacola, FL: Naval Education and Training Command.

Vaden, E. A., and Hall, S. 2005. The effect of simulator platform motion on pilot training transfer: A meta-analysis. *International Journal of Aviation Psychology* 15:375–393.

Wiegmann, D., and Shappell, S. 1999. Human error and crew resource management failures in naval aviation mishaps: A review of U.S. Naval Safety Center data, 1990–96. *Aviation, Space, and Environmental Medicine* 70:1147–1151.

Wilson-Donnelly, K. A., and Shappell, S. A. 2004. U.S. Navy/Marine Corps CRM training: Separating theory from reality. *Proceedings of the Human Factors and Ergonomics Society 48th Annual Meeting*, New Orleans, LA. 2070–2074.

Wright, R. H. 1973. Retention of flying skills and refresher training requirements: Effects of non-flying and proficiency flying (HumRRO technical report 73-32). Alexandria, VA: Human Resources Research Organization.

6 Stress, Human Reactions, and Performance

6.1 INTRODUCTION

An important part of psychology is the study of variations in how we think, feel, and react. Although it is important to be aware of such variations, there are a number of commonly shared patterns in terms of reactions—for example, to dramatic events. Hence, this chapter discusses both individual differences and common traits in reactions to everyday stress and more significant incidents. Also, this chapter investigates common psychological reactions in passengers.

6.2 PERSONALITY

Personality is a sweeping construct. It may be defined broadly as every internal factor that contributes to consistent behavior in different situations or, narrowly, as encompassing only emotions and motivation. A broader definition of personality may include intelligence; traditionally, however, personality has been considered separate from intelligence and skills. This distinction is evident in psychological tests, which are usually divided into ability tests and personality tests. Ability tests often include time constraints, and the objective is to get as many correct answers as possible, whereas personality tests seek to measure typical response patterns—that is, how an individual usually reacts to a given situation.

For a long time, the psychology community has been engaged in discussion on how many personality traits or dimensions are necessary to describe someone. Imagine describing a long-time friend. Which adjectives and examples should be used? Perhaps the words "great," "friendly," "humorous," or "reliable" spring to mind. If one were to describe a person of whom one was less fond, perhaps words such as "aggressive," "cynical," "egotistical," or "prejudicial" would be used. If such descriptions are collected and systematized using factor analysis (see Chapter 2), five general categories emerge. These five factors are normally referred to as "the big five": *extroversion, agreeableness, conscientiousness, neuroticism,* and *openness to experience* (Costa and McCrae 1997). Some measures use *emotional stability* instead of neuroticism. In other words, the positive end of the scale is applied. Refer to Table 6.1 for examples of characteristics that lead to high or low scores for the different dimensions.

Most techniques for personality characteristic measurements use statements combined with a point scale ranging from one to five (or, in some cases, one to seven) to which subjects note their level of agreement. Combinations of positively

TABLE 6.1

Overview of Traits Included in the Five-Factor Personality Model

Characteristics/Traits	Characteristics of Individuals with Low Scores	Characteristics of Individuals with High Scores
Extroversion	Passive, quiet, introverted, reserved	Open, talkative, energetic, enjoys social situations
Agreeableness	Cold, cynical, unpleasant, directly expresses aggression	Gentle/kind, cooperative, avoids conflict, credible
Conscientiousness	Unreliable, disorganized, irritable, prefers not having plans	Conscientious, responsible, organized, goal oriented
Neuroticism	Calm, not neurotic, comfortable, deals with stress	Uneasy, worried, nervous, emotional
Openness to experience	Conventional, practical, down-to-earth	Intellectual, cultural, open to new experiences

and negatively phrased statements are often used for the different dimensions. For example, "I am often anxious" may be used instead of "I am never anxious."

Factor analysis of adjectives (e.g., "kind," "friendly," "firm," and "anxious") and longer statements has been the starting point for the five-factor model. It is also possible to break the five main factors into subfacets if a more detailed description of the person is required. Studies have shown that this five-factor solution may be replicated across language and cultural barriers, and a satisfactory correspondence between the subjects' description of themselves and how others perceive them has been established, particularly when described by persons who know them well (see, for example, Digman 1990).

However, not all researchers agree that the five-factor model represents a comprehensive description of personality. Some think it contains too few (or too many) traits. Others find the model simplistic or that it fails to explain "how we have developed into being who we are" (refer to Block, 1995, for a critical analysis). Despite these criticisms, the model has been widely used in research involving personality and appears to be accepted widely as a good starting point for personality assessments (see, for example, Digman 1990; Goldberg 1993). Studies have demonstrated a considerable inheritable component in personality characteristics and that personality traits develop until the age of 30, after which they remain relatively stable (Terracciano, Costa, and McCrae 2006).

In the 1980s a series of personality inventories were developed to map the big five traits. Costa and McCrae's *NEO personality inventory* (NEO-PI) (1985) is particularly well known. The five traits are presumed to be relatively independent of abilities; however, one exception is "openness to experience," which, to a certain extent, correlates with intelligence (Costa and McCrae 1985). These approaches share the advantage of a generally high reliability level. In terms of predictive validity, several of the dimensions have proven to be associated with work achievements, although the correlations are described as small or moderate (see Chapter 4). A meta-analysis of the relation between the personality traits and accidents revealed conscientiousness and agreeableness to be associated systematically with accident involvement: People with low conscientiousness and agreeableness scores had experienced a greater number of accidents than those with higher scores (Clarke and Robertson 2005).

Although these measurement techniques were designed to measure aspects of an individual's established personality, other methods or diagnostics systems are used to document problems, such as high levels of anxiety and depression, or to establish whether someone is suffering from mental illness.

In addition to the five-factor model and its associated empirical systems, specific traits are often used to describe personalities or aspects thereof that may be of importance in some situations. These traits are typically tied to certain theories or are particularly suited to explain reactions (or predict behavior) in certain situations. The following are examples of such traits: *type A behavior, locus of control (LOC), psychological resilience,* and *social intelligence.* Type A behavior and LOC are discussed in detail in Section 6.6 on individual differences and stress.

Psychological resilience has been studied in particular in relation to people who thrive despite challenges and misfortune. Important factors here are personal attributes such as social skills and leading a structured life; however, external support from family and friends is also important (Friborg et al. 2005). Social intelligence usually refers to social skills and the ability to understand one's own and others' reactions (Silvera, Martinussen, and Dahl 2001). These traits are more or less related to the personality traits contained in the five-factor model. For example, LOC correlates with neuroticism, such that those with internal LOC are considered more emotionally stable. In addition to traits mentioned in scientific literature, a number of poorly documented theories on personality can be found in popular science magazines and the like, often accompanied by different tests that reportedly measure these theories' accompanying traits. However, documentation to support the reliability and validity of such tests is usually scant.

6.3 WHAT IS STRESS?

We are continually bombarded with influences, expectations, and demands placed on us by our surroundings. Work commitments or the lack of time and resources to complete tasks are typical examples. Both paid work and unpaid work (e.g., caring for family members) are applicable factors in this regard. To meet social demands or solve work-related tasks, the individual relies on different sets of resources, including knowledge, experience, and personal attributes. Some theories describe stress as the result of factors or elements that have a negative impact on the individual—for example, distracting noise or pressure at work (stimulus-based theories). Other theories are concerned with the consequences of stress, such as various emotional and physical reactions (response-based theories). The latter tradition is exemplified by Selye (1978). He describes a general stress response that is valid for everyone and consists of three phases: the alarm phase, the resistance phase, and the exhaustion phase.

A more modern understanding requires stress to be regarded as the interaction between demands and the resources available to the individual. When demands placed on an individual exceed his or her resources, stress develops. In these interaction models, an important point is that the person must evaluate the demands and consider whether these demands exceed his or her resources. Due to this cognitive evaluation, what one individual considers a stressor is not necessarily considered a stressor by someone else (noted in, for example, Lazarus and Folkman 1984). Balance between external demands and personal attributes is perceived as challenging and satisfying to the individual (Frankenhaeuser 1991), whereas imbalance is a precursor to emotional, physical, and behavioral consequences.

Frankenhaeuser's (1991) *bio-psychosocial model* (depicted in Figure 6.1) delineates the relationship between stress and health. In this model, the person is subjected to various demands, such as intense workloads, time constraints, shift work, problems, or conflicts. The person relates this to his or her resources, including experience, physical and mental health, personal abilities, and, potentially, external support. If demand surpasses the person's resources, stress ensues, accompanied by both psychological and physiological reactions. Immediately, various stress hormones are released into the body (adrenaline, noradrenaline, and cortisol). These hormones produce a number of advantageous effects in precarious situations; however, problems may arise if the individual is exposed to these effects for an extended period of time. If a person is continually stressed or if there is not enough time to rest, the body is unable to normalize the physiological reactions in time for the next work session. Stress is also an unpleasant experience, with

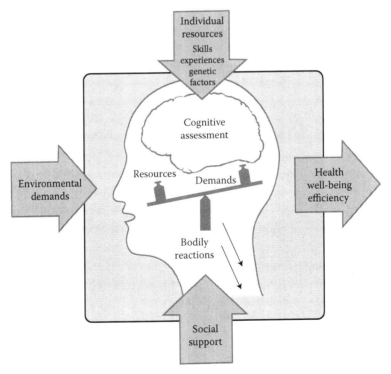

FIGURE 6.1   The Frankenhaeuser (1991) bio-psychosocial stress model. (Reproduced with the kind permission of Springer Science and Business Media.)

short-term and long-term consequences for the affected person's productivity. We will discuss this more in a later section.

Other models describe work-related stress, such as Karasek's demand-control model (Karasek 1979; Karasek and Theorell 1990), which describes how stress relates to various consequences such as health risks and behavior within the organization. In this model, work-related demands are described as "high" or "low"; similarly, the individual's ability to affect or control the situation is deemed "high" or "low." Combining high demands with low levels of control increases the risk of psychological impacts and physical illness, such as cardiovascular diseases (Yoshimasu 2001). On the other hand, combining high demands and a high level of control encourages learning and has a motivational effect. Later expansions on this model have pointed out that social support, such as assistance and encouragement by colleagues, may reduce stress and minimize risks associated with negative consequences of stress. There are several forms of social support, such as care and empathy, as well as assistance of a more practical nature such as being applauded for doing a good job.

6.4 CONFLICTS BETWEEN WORK AND PRIVATE LIFE

Today, many of us choose to combine work and private life. This means that many people must be adept at several roles (parent, partner, employee, and so forth). The total workload is substantial and may lead to insufficient time for recreation and rest. Thus, conflicts may arise from the interaction between work and private life. At the same time, having multiple roles can have positive aspects, such as increased self-confidence and greater financial freedom. There are several approaches to work-to-home conflicts. One is that time management becomes difficult, and it seems like "there are not enough hours in the day." Another is that work causes stress and exhaustion, leading to the inability to engage in quality family time as much as one would like.

Several studies have described a connection between work–home conflicts and burnout (Martinussen and Richardsen 2006; Martinussen, Richardsen, and Burke 2007) and between work–home conflicts and reduced satisfaction with one's partnership, as well as reduced job satisfaction (Allen et al. 2000). Some studies have pointed to a so-called "crossover effect" between partners: Stress and tension experienced at work by one person are transferred to his or her partner, who subsequently has to deal with the stress by serving as a buffer (Westman and Etzion 1995). This transfer probably occurs because the person empathizes with his or her partner; however, a more direct effect is plausible because exhausted and frustrated persons have "less to give" when they come home from work. A study of couples with young children revealed that men were more likely to become passive and withdrawn upon returning home after a difficult day at work, and women were more likely to become aggressive (Schultz et al. 2004). In short, the study indicated that having a difficult day at work might have consequences for one's partner and that there are gender-related differences in how one reacts to such situations. These findings have since been supported by a survey of male flight controllers that revealed that they often reacted with withdrawal after a stressful day at work. The study also revealed that satisfaction was greater when their partner accepted such behavior.

Although fewer studies have investigated how family or private matters negatively affect job performance, we can safely assume that such effects would be undesirable. Some examples of demanding tasks at home include dealing with disease, partnership breakdowns, and caring for many young children. However, negative emotions are not the only emotions transferable between home and work. Positive experiences at work may transfer to family life as one arrives home contented and uplifted; conversely, positive events at home may lead to a better day at work. Arguably, people with family commitments may find it easier to set boundaries for their work commitments, which, in the absence of a family, might have absorbed a greater part of the day. Thus, family commitments become a legitimate excuse to the employer and, not least, to oneself. In particular, young, single professionals presumably experience greater pressure to perform and are more likely to work longer hours, indicating that individuals with family commitments are not the only ones struggling to find the balance between work and leisure. Further, modern communications (such as e-mail and cell phones) may enable a person to continue working even after official work hours.

Several studies have indicated the continuation of traditional labor-sharing practices in households in which women account for cooking, cleaning, and caring for children and men are mainly responsible for tasks such as maintenance and car repairs (Lundberg and Frankenhaeuser 1999; Lundberg, Mårdberg, and Frankenhaeuser 1994; Østlyngen et al. 2003). A Norwegian study involving parents with young children (0–6 years) demonstrated that females did about 70% of the domestic work; however, this study included a relatively high percentage of mothers who worked part-time (Kitterød 2005). Nonetheless, the study indicates a larger total workload for women in comparison to men, leaving them with a reduced amount of time (after finishing the day's paid and domestic work) available for relaxation and recreation.

A study of junior managers employed at the car manufacturer Volvo in Sweden revealed that stress hormone levels were equal in female and male managers during work hours, but a difference was noticeable after work hours. Rising levels were recorded in females between 6:00 and 8:00 p.m., but in males the corresponding values decreased during the same period (Frankenhaeuser 1991). The physiological data were consistent with self-reporting of weariness. Thus, male junior managers started relaxing immediately after work. This was not the case for females until much later in the evening. Therefore, females had a shorter time available for relaxation than did males, possibly incurring negative health consequences in the long term (Lundberg 2005).

6.5 BURNOUT AND ENGAGEMENT

Burnout can be regarded as a stress reaction occurring after long-term, work-related demands and pressure. Maslach and Jackson (1981, 1986) have defined burnout as a three-dimensional psychological syndrome consisting of *emotional exhaustion* (a condition including overwhelming emotional and physical exhaustion), *depersonalization* (characterized by negative emotions and cynical attitudes toward the recipients of one's service or care), and reduced *personal accomplishment* (a tendency to evaluate one's own work negatively). Initial studies of burnout were based on workers in care professions such as nursing. Then, burnout was considered to be triggered by high interpersonal demands. More recently, however, burnout has been found in professions that do not necessarily involve caring for patients, clients, or pupils. The three dimensions have subsequently been generalized into the dimensions of *exhaustion, cynicism,* and *professional efficacy* (Richardsen and Martinussen 2006).

A number of work environment factors have proven to be associated with burnout. Leiter and Maslach (2005) have described six categories of such factors. One of these is workload: too much to do (or not enough time available to do it) or insufficient resources to solve given tasks. Insufficient control or autonomy in the workplace is also associated with burnout. If a person feels powerless to control resources required to complete tasks or unable to influence how the job is done, reduced personal accomplishment may result.

Some work-related factors have a preventive effect on burnout or serve as a "buffer." These factors include social support or assistance provided by management or colleagues. Being rewarded, acknowledged, and feeling fairly treated (in terms of promotions, etc.) are positive resources. Sometimes an employee may find that the organization has values different from his or hers—for example, being told to withhold information or deceive someone. At present, however, what happens when values held by employees differ from those of employers has been insufficiently researched.

Although early research into burnout was directed at particular jobs (such as the previously mentioned nurses), not only those in demanding care professions are at risk. Few studies have been aimed at burnout in aviation professions, and most work-related stress studies focus on short-term effects, such as measuring blood pressure changes in flight controllers when exposed to elevated air traffic density levels.

A study of Norwegian air traffic controllers concluded that the burnout rate was not significantly higher for them than for other professions included in the study. Both the levels of conflict experienced and work–home-related conflicts were associated with exhaustion in this group (Martinussen and Richardsen 2006). Most people would consider flight control to be a highly stressful profession; thus, it is surprising that this group did not have elevated burnout rates. On the other hand, flight controllers go through a process of strict selection, education, and training to enable them to perform extremely demanding tasks. Hence, there appears to be a balance between the tasks to be solved and the skills and abilities of the employees. This does not mean that the individuals are immune to the demands and requirements of the organization or that access to resources would not have a positive effect. Similar results have been found for the police profession, which considers organizational issues to be more frustrating and demanding than the job itself (Martinussen et al. 2007).

Traditionally, studies into work environments and burnout have focused greatly on negative aspects and attempted to establish the illness-inducing sides of the work environment (Maslach, Schaufeli, and Leiter 2001). Recently, however, researchers have turned their eyes on studying the opposite of burnout—that is, engagement—to find out what causes this outcome. Schaufeli and Bakker (2004) present a model that describes how resources and workplace demands relate to both engagement and burnout (Figure 6.2). This model tells us that burnout is, first and foremost,

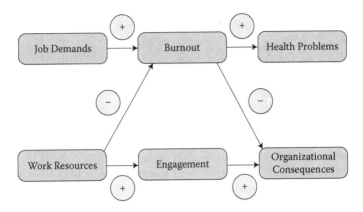

FIGURE 6.2 The job demands–resources model for burnout and engagement.

associated with demands and, second, with a lack of resources; engagement is predominantly associated with access to resources such as rewards, recognition, and support. Burnout and engagement have consequences for the organization. Burnout implies negative consequences, whereas engagement has a positive impact. Examples of organizational consequences are intention to quit, work satisfaction, work performance, and feeling commitment to the organization. Burnout also has negative consequences for the affected individual's health and quality of life.

6.6 INDIVIDUAL DIFFERENCES AND STRESS

There is little doubt that some working environments or factors are generally considered stressful. Yet, people with certain personality characteristics experience stress more often and more intensely than others. It is therefore of interest to study these differences in detail. Studies have shown that neuroticism is associated with burnout. Although some of the other big-five personality characteristics also have been found to be associated with burnout, findings vary greatly between studies.

Working with individuals suffering from cardiovascular disease, two physicians (Rosenman and Friedman 1974) claimed to have observed certain recurring traits in their patients. They labeled some of their patients' behavior *type A*—characterized by irritability, time constraints, competitive mentality, aggression, hostility, and ambition. Patients who did not display these properties were labeled *type B*. Several methods of measuring and mapping type A personalities have since been developed, of which the most well known and widely used is the Jenkins activity survey. This method condenses the mentioned characteristics into two dimensions: *impatience–irritability* and *achievement strivings*. The latter represents a positive side—namely, that the person sets goals and works hard to achieve them. The former dimension represents the less positive side and is characterized by impatience and aggression. Impatience–irritability is related to health issues, while achievement strivings are associated with enhanced work achievements and better student grades.

Researchers have established the existence of a significant inheritable component in type A behavior; type A has also been found to be a risk factor in the development of cardiovascular diseases (see, for example, Yoshimasu 2001). The link is thought to occur because type A—in particular due to the impatience–irritability aspect—is associated with greater physiological activation including elevated blood pressure levels and heart rate. However, lifestyle choices, such as alcohol and smoking, may also contribute to this pattern.

What are the consequences of type A behavior in the workforce? In fact, many organizations are likely to reward type A personalities, at least those aspects relating to striving for achievement; however, irritability may cause problems for the individual and his or her surroundings. Perhaps type A behavior directly influences work-related satisfaction and the experience of stress and burnout. The combination of aspects of the work environment and, for example, impatience–irritability may cause particularly unfortunate outcomes for certain individuals. Type A personalities may even intuitively choose professions that are more challenging or involve a greater pace and workload. It may be the case that type A personalities affect their work environment in certain ways and therefore contribute to creating a more stressed environment.

Another personal attribute under investigation is locus of control—the extent to which the person feels that he or she can influence or control events and situations. Internal LOC has been shown to be associated with several work-related variables, including improved motivation and commitment (Ng, Sorensen, and Eby 2006). People with active coping techniques (i.e., who act strategically to handle difficult situations) generally score lower on burnout than people who use more emotion-focused coping techniques (e.g., the person attempts to deal with emotions by seeking comfort in other people).

People also differ according to gender, age, and other circumstantial factors. Generally, low correlations exist between burnout and factors such as age and gender. Sometimes, younger professionals have been found to be more exposed to burnout, but in other studies (e.g., flight controllers and the police), the "age effect" is reversed (Martinussen and Richardsen 2006; Martinussen et al. 2007). However, such correlations are found to be weak. This also applies to differences due to gender: Some studies find that females report a greater degree of exhaustion than men and that men have higher cynicism scores; others find no gender differences.

6.7 CONSEQUENCES OF STRESS

Stress has both short-term and long-term consequences. The following discusses emergency situations and how this affects an individual in relation to job performance. A critical stress situation may arise when something unusual takes place during a flight. Examples of stress situations include indications of technical difficulties or a rapid deterioration of weather conditions. It is important to know typical reactions in such situations, be aware of how the crew responds to stress, and understand how it affects decisions made during the flight.

It is difficult to measure how stress affects various cognitive functions. Ethical considerations are involved in exposing subjects to stressful situations in experiments to study how they react. A possible solution to this problem is to use a simulator to study how people handle various abnormal situations. This provides a controlled environment in which emergency situations can be manufactured, workloads and time constraints increased, and reactions recorded. Although many simulators are highly realistic, they will never be identical to real-world experiences; thus, the possibility that findings cannot be transferred to real-world situations must not be ignored.

A second option is to study incidents that have already taken place, reconstruct the decisions made, and observe how stress contributed to the event. The drawback of this method is that it is based on human recollection and perception of the event and what happened. Another issue is that situations may appear quite differently in hindsight compared to how the situation actually occurred and was experienced by the people involved. No matter which design is chosen, there are challenges and shortcomings; however, some findings on the immediate effect of stress on cognitive functions and decision-making capacity are consistent.

According to a review provided by Orasanu (1997), stress may have the following effects:

- People make more errors.
- Attention is reduced, causing tunnel vision or selective hearing.
- Scanning (vision) becomes more chaotic.
- Short-term memory is reduced.
- Change of strategy: speed gains preference to accuracy. People act as though time limits apply. Strategies are simplified.

Thus, cognitive functions are subject to a number of stress-related consequences in terms of how we perceive our surroundings, process information, and make decisions. An important aspect regarding aviation is the need to take in and monitor information constantly. In a high-stress situation, the capacity to do so diminishes, reducing the ability to understand what is being said over the radio or what another person says. Similarly, the information that is supplied to someone stressed may be poorly understood or not understood at all.

By short-term memory, we refer to the processes or structures that contribute to the temporary storage and processing of information. It enables us to read a couple of sentences while storing and processing information about, say, the last word in each sentence and to repeat those words in the correct order. Some argue that there are clear limitations to short-term memory, and early studies revealed a short-term memory capacity in adults of seven numbers (±2). However, later studies have pointed out more of a complexity in this matter. For instance, grouping a set of numbers and remembering them as one (e.g., "2," "4," and "5" as "245") makes it possible to recall even more numbers. On the other hand, more complex elements (e.g., words) are more difficult to remember, reducing the number of retainable elements to less than seven. In stressful situations, this capacity will be further reduced, with consequences for the ability to perform basic mental arithmetic, such as calculating the remaining flight time when the fuel tank is half full.

In connection with incidents and accidents, much focus has been put on decision making: How was the situation perceived, and what was the chosen course of action? Klein (1995) has studied decision making in real-life situations, in aviation and in general, for many years, and he developed a model called *recognition-primed decision* (RPD). His model involves experts using their knowledge to recognize or identify the problem and choosing a solution that has been proven successful in previous, similar situations. If the solution to the current situation is suitable, it is applied, and only one solution is considered at a time.

This model has since been expanded by Judith Orasanu (1997), who describes a model that can be used even in unfamiliar situations. The first thing the person does is to evaluate the situation: What is the problem? Are the warning signs clear and unambiguous or are they changing? Often, experts in such situations also consider how much time is available to take the necessary steps and they evaluate the level of risk involved. Then, the person must evaluate whether any existing procedure is available to remedy the situation. Perhaps there is more than one solution? One possibility is, of course, that there are no known solutions, which triggers the need to create a new and untested course of action.

In general, cognitive processes that involve retrieval of information from long-term memory are resilient to stress, while processes that require the use of short-term memory are more vulnerable. In other words, if the warning signs are well known and clear and a standard solution is applicable, the situation will not be significantly prone to stress. On the other hand, if the signs are obscure or keep changing and multiple courses of action must be considered, the situation is prone to stress. Perhaps it is not surprising that experienced pilots do not make as many mistakes under pressure as less experienced pilots: They have a greater number of experiences stored in their long-term memory and are more likely to use a rule-based approach rather than having to consider several options or even improvise new solutions.

It is therefore important to have knowledge of stress and how we are affected by acute stress; for example, a common mistake is to assume that time is more precious than it is. Another consequence of stress is the simplification of strategies, such as preferring speed over accuracy. It is important to be trained in managing stressful situations to familiarize oneself with critical situations and learn how they can be resolved. It is also important to be aware of strategies to reduce workloads in stressful situations—for example, how to distribute tasks between members of the crew in the best possible way.

6.8   SHIFT WORK

Shift work is commonplace for many people in aviation, as are night shifts for some workers. Crew members who travel across time zones also experience problems due to jet lag. The notion of shift work usually applies to work taking place outside regular daytime hours (6 a.m.–6 p.m.); however, there are different types of shift work arrangements, many of which apply some form of rotating shifts. Shift work has several consequences in relation to health issues, sleep, work achievements, the risk of accidents, increased work–home conflicts, and participation in social activities taking place on weekday evenings and weekends. Working shifts may also influence one's relationship with the employer—for example, in the form of reduced work satisfaction (Demerouti et al. 2004).

6.8.1 SLEEP

Sleep difficulties are one of the most common problems associated with shift work in general and night shifts in particular. In human beings, body temperature and the production of hormones, stomach acids, and urine follow a cycle of approximately 24 hours. External factors such as light exposure influence the internal clock (or "body clock") and its adjustment. Hence, at night, the body is expecting to do something completely different from working. Most adults usually sleep between 6 and 9 hours each night, averaging from 7 to 7.5 hours (Ursin 1996). Physiological measurements make it possible to map the various phases of sleep, including the *rapid eye movement* (REM) phase, in which dreams occur. Throughout the night, the different phases are repeated. Toward the end of the night, deep sleep subsides and it is common to wake up a couple of times. Sleep varies according to body temperature, which is at its lowest point in the morning (between 4 and 6 a.m.) and this is when it is most difficult to stay awake (Pallesen 2006).

Several theories have been proposed on the function of sleep—that is, why we sleep. Some are based on the theory of evolution and postulate that it is safer to be inactive in darkness because we cannot see where we are going. A second group of theories argues that sleep has an important restorative function for the body and that certain types of hormones (which boost the growth of body tissues) are produced during sleep (Pallesen 2006). In addition, it appears that sleep has a restorative function relating to brain cells and their protection from cell degeneration processes (Pallesen 2006).

Sleep is normally regulated by how long we have been awake as well as our daily routines and habits (Waage, Pallesen, and Bjorvatn 2006). Thus, it is more difficult to sleep during the day than the night; people who go to bed after a night shift commonly experience less total sleep and more frequent sleep interruptions, and they may have to get up to go to the toilet even though this normally is not necessary when sleeping through the night. There are, however, individual differences in tolerance to night and shift work. Some studies have shown that mature individuals (over the age of 45) have greater problems sleeping or resting after working a night shift and that problems increase with age (Costa 2003). Some people, however, find it easier to work night shifts as they get older. Increased experience in this type of work and the acquisition of adaptive techniques are possible interpretations of the results in these cases. Perhaps the worker's children have grown up, making the family situation more accepting of sleeping during the day.

Typically, however, studies into the consequences of shift and night work are likely to be influenced by a "selection effect." It is reasonable to assume that those who experience great discomfort with such work hours would quit the job after a period of time, resulting in a selection process in which those who experience less discomfort continue working. Estimates show that as many as 20% of shift workers quit after a relatively short period of time (Costa 1996).

6.8.2 HEALTH IMPLICATIONS

Several studies have found adverse health implications in shift workers, including increased exposure to cardiovascular diseases, problems with digestion, and cancer

(Costa 1996). A longitudinal Danish study that monitored a vast number of subjects over 12 years demonstrated that shift workers were more likely to develop cardiovascular disease compared to daytime workers (Tüchsen, Hannerz, and Burr 2007). Females working shifts more frequently report problems linked to menstruation; some studies have found a connection between shift work and miscarriages, low birth weight, and premature birth (Knutsson 2003). A meta-analysis of 13 studies on the link between breast cancer in women and night shift work suggested an increased breast cancer risk. Approximately half of these studies were based on cabin crew, while the remaining studies were based on females in other types of night shift work. The reason for the elevated cancer risk is uncertain, but a possible explanation is that working nights reduces the production of melatonin, which is considered to have a cancer-preventing effect (Megdal et al. 2005).

Presumably, the negative health effects are direct consequences of disruptions to the biological 24-hour rhythm, as well as behavioral changes due to shift work. For example, sleep deprivation can lead to the use of alcohol to induce sleep or excessive smoking to stay awake at night. It is also possible that shift work contributes to work–home conflicts, which heighten stress levels and further exacerbate the adverse health implications of shift work.

6.8.3 ACCIDENT RISK

A number of studies have examined how sleep deprivation affects performance and how it relates to accidents. Laboratory studies have looked at the connection between reduced sleep and cognitive and psychomotor tasks, revealing that tasks requiring constant attention are more affected by sleep deprivation; more advanced tasks, such as reasoning, were less affected (a summary can be found in Åkerstedt 2007). One study, in which subjects were asked to operate a flight simulator at night, demonstrated that their reduction in performance was equivalent to a blood alcohol level of 0.05% (Klein, Brüner, and Holtman 1970).

The consequences of sleep deprivation generally increase during an extended period without sleep. However, a person's day-to-day routines are also important; performance improves during the day relative to the night (Åkerstedt 2007). Even after a prolonged period of being awake, performance will improve somewhat during the time period in which an individual is normally awake (the "daytime" according to the person's body clock).

Working shifts and working night shifts in particular are associated with elevated risk of accidents (a summary is provided in Folkard and Tucker 2003). Several studies from the field of medicine show that doctors make more mistakes if they are sleep deprived (e.g., during 24-hour shifts) and need longer time to perform basic tasks, such as intubating a patient (Åkerstedt 2007). A number of studies into motor vehicle accidents point to drowsiness—particularly driving after working night shifts—as an important factor in many accidents. In a survey of pilots who were asked to describe how drowsiness typically affected them, the most common symptoms were reduced attention and lack of concentration (Bourgeois-Bougrine et al. 2003). The remainder of the crew reported evidence of tiredness, primarily in the form of longer response times, greater error frequency, and poor communication (Bourgeois-Bougrine et al. 2003).

6.8.4 PRIVATE LIFE

There are individual differences in the ability to deal with shift work. These differences are usually attributed to biological, social, and health issues, although the organization of the work and the shift schedule do play a part. Employees who are privileged enough to have a certain level of flexibility and influence on the shift work roster are generally more content and experience fewer problems associated with shift work (Costa, Sartori, and Åkerstedt 2006).

For most people, however, shift work is problematic and has significant consequences for private life. There are issues with meeting family commitments and participating in social activities, which typically take place in the afternoon. Having a day off on a weekday is not the same as having a day off on the weekend. Family commitments may also make it more difficult to sleep during the day to recuperate after a night shift or having been awake for a long period of time. Mood volatility after night shifts undeniably places additional demands on the individual and the family. A study of police officers in The Netherlands concluded that the timing of shift work is a decisive factor as to the level of work–home conflicts and recommended avoiding shifts that involve regular weekend work (Demerouti et al. 2004).

Some studies have described problems with combining shift work and family life as higher for females than males. In particular, women with young children report shorter and more interrupted sleep after night shifts, as well as accumulated tiredness (Costa 1996).

6.8.5 JET LAG

During eastbound and westbound long-haul flights, several time zones may be crossed. Longer trips generate greater divergence from the biological clock, as well as longer work hours for pilots and crew. Common symptoms of jet lag include problems sleeping at designated times, daytime weariness, concentration and motivational issues, reduced cognitive and physical abilities, headaches, and irritability. Reduced appetite and digestion problems may also occur.

It takes time to adapt to a new time zone. As a rule of thumb, resetting the biological clock takes about 1 day per time zone crossed or entered. Thus, on a flight from Oslo to New York (a 6-hour time difference), it would take 6 days to adjust. However, crew members are often stationed for a shorter period than is necessary to come to terms with the new routine; the stay is interrupted by the return flight or perhaps another flight to a new time zone. Jet lag is slightly less severe on westbound flights compared to eastbound flights. This is because the body finds it easier to adapt to a slightly longer day (when traveling west) than a slightly shorter one (traveling east) because the biological clock is slightly longer than 24 hours for most people. It is easier to stay up late and postpone sleep than forcing oneself to sleep earlier than the normal time.

Exposure to light in certain periods may ease the impact of a new environment. Using the drug melatonin to fight the effects of jet lag is controversial for air personnel and, in fact, discouraged. First, evidence for its effectiveness is insufficient and, second, negative side effects such as reduced attention span during work hours have not been ruled out (Nicholson 2006).

6.8.6   How Can One Prevent Drowsiness?

When a person is required to stay awake for long periods or during night shifts, performance deteriorates and the probability of errors increases. Particularly at times when there is not much to do (e.g., on long-haul flights) staying awake may become a problem. There are various countermeasures to sleepiness, such as having a power nap (short nap) or engaging in some physical activity. This may not always be possible due to the nature of the job. Some substances affect both wakefulness and performance; caffeine is a well-known example. If possible, performing a different type of task for a while can help. Some people find it worthwhile to take a nap before a night shift, but others find it difficult to sleep in the afternoon, perhaps because of family commitments such as taking children to soccer practice. Generally, it is important to sleep well between shifts (good sleep hygiene).

6.9   EXTREME STRESS

Persons affected by accidents in aviation, both directly and indirectly (e.g., close colleagues perishing in a plane crash), are exposed to stress that is different from everyday stress. Extreme stress reactions vary from a strong feeling of surreality immediately after the event to an apparent absence of a reaction. However, the long-term effects must also be considered; although an individual may seem to handle the situation well initially, it may take time for a reaction to manifest itself.

Persons exposed to trauma are at risk of developing posttraumatic stress disorder (PTSD), characterized by discomforting thoughts or dreams in which the trauma is relived. Affected individuals feel numb and avoid situations that remind them of the accident. Restlessness, nervousness, and sleep problems are also common. Posttraumatic stress disorder can lead to a reduction in the affected individual's ability to perform duties to the same level as before the accident. In some cases, the condition is characterized by fear, helplessness, aggression, and/or a hostile attitude. Why the condition is characterized by negative emotions has not yet been made clear; however, a possible explanation is that people who have been exposed to a traumatic event have a reduced threshold for perceiving a situation as threatening. This leads to anxiety and avoidance, but also to aggression and more easily finding oneself in an attack position.

The link between the severity of PTSD symptoms and aggression is stronger in individuals traumatized by acts of war than in individuals experiencing other traumatic events (Orth and Wieland 2006). In most cases, these symptoms will subside in the weeks and months following the experience, although a few individuals develop a chronic condition associated with depression, substance abuse, anxiety, and inability to work. Even though most people regain their ability to work after a traumatic incident, it is important to be able to identify individuals who are at risk of developing PTSD. It is safe to assume that social interventions, such as care and support from colleagues and management, will be of use to many of those affected. Population studies from the United States have shown that 50–60% of the population has experienced a traumatic event at least once in their lifetime, but only a small proportion (5–10%) developed PTSD (Ozer et al. 2003). Therefore, it is of interest to explore the factors or conditions that help us understand why some people develop PTSD and others do not.

A meta-analysis of a number of studies examining these factors found that individual factors, such as a record of previous mental health problems and prior exposure to trauma, were associated with more severe PTSD symptoms. Other factors, such as previous involvement in life-threatening situations, social support networks, and which emotions were experienced during the event itself were also associated with PTSD symptoms (Ozer et al. 2003).

With regard to vulnerability factors, those who were directly involved in the accident are considered more vulnerable than those not directly exposed to the event. Persons not appropriately trained—for example, in relation to the necessary rescue efforts—are more likely to have stronger and longer lasting reactions. In addition, the severity of the situation (e.g., as measured by the number of dead and injured or intense and overwhelming sensory impressions) is likely to increase the probability of developing PTSD.

All organizations should have routines for how to deal with accidents and how to care for those involved after the accident. The most common approach is to review the event with those involved within 48–72 hours after the event. This requires getting together in groups under the leadership of someone trained in such exercises. Typically, this entails restating exactly what has happened as well as sharing thoughts and emotions. Those involved are informed about common reactions to accidents or dramatic events. These measures are intended to minimize acute symptoms and make those involved more capable of dealing with their reactions in the time to come.

Another benefit of conducting such group interventions is the identification of people in need of extra counseling and care. For those who need more support, there are a number of different types of individual therapy, such as cognitive–behavioral therapy, which contains an exposure part and provides a structured framework to deal with thoughts and emotions. Other cognitive–behavioral methods emphasize assisting the person in dealing with anxiety. A different form of therapy consists of what is called "eye movement desensitization," which consists of having the patient visualizing the traumatic event while watching a moving stimulus.

Meta-analyses have found that therapy represents an efficient treatment of PTSD and that up to 67% of those who complete the treatment no longer satisfy the criteria to be rediagnosed with PTSD (Bradley et al. 2005). It is important to encourage affected individuals to keep working and maintain frequent exposure to the situation—for example, by returning to aviation. Studies have revealed that those who return to work are better off in the long term than those who choose to change professions, even for individuals who initially had similar symptoms.

6.10   PASSENGER REACTIONS

Most of the intended readers of this book love to fly, but others consider air travel best avoided. Some people become argumentative or quarrelsome, risking a fine or imprisonment for their behavior. The final part of this chapter will therefore deal with distressed and disorderly passengers.

6.10.1   Fear of Flying

Extreme fear of flying, or flight phobia, implies an exaggerated fear of flying compared to the real risk involved. Affected individuals travel by air in great discomfort or avoid it altogether. Some have to travel by air as a work requirement, in which case fear of flying can be particularly inconvenient. Fear of flying can also be a significant impediment to holiday arrangements. In a survey of a group of randomly selected Norwegian nationals, about one in two said he or she was never afraid of flying, while the remainder indicating varying degrees of discomfort (Ekeberg, Seeberg, and Ellertsen 1989). International studies indicate that between 10 and 40% of passengers are affected by fear of flying (the significant variation in this figure is probably due to phrasing of questions and the composition of the group surveyed). A Norwegian survey (Martinussen, Gundersen, and Pedersen 2008) asked participants: "To which degree are you afraid of flying?" The results are outlined in Table 6.2.

   The group surveyed consisted of students as well as passengers surveyed at an airport in northern Norway. The survey found no correlation between age and fear of flying, but it did find that females were more likely to report fear of flying than males. Participants disclosing any form of discomfort during flight were asked about which factors caused the greatest amount of fear; commonly, these were cabin movement, vibrations, noises, or announcements of turbulence. Participants were also asked how they reacted to such situations. Typical reactions were experiencing palpitations and believing something was going to go wrong. In addition, many

TABLE 6.2
Extent to Which the Survey's
Participants *Are Afraid of Flying*

	Total (N = 268)
Not afraid at all	56%
Sometimes a little afraid	29%
Always a little afraid	8%
Sometimes very afraid	4%
Always very afraid[a]	4%

[a]   Combined category based on three alternatives, all involving very afraid.

reported great attentiveness to aircraft noises and closely watching the behavior of cabin crew.

Even aviation personnel can develop a flight phobia. Although the prevalence of this problem is unknown, one study examined a group of aviation personnel who were seeking support for various psychological problems. Fourteen percent reported fear of flying, a quarter of whom said they had experienced an accident or knew someone who had been involved in one. In addition to fear of flying, about half of the individuals in the group had been diagnosed with a mental illness such as depression (Medialdea and Tejada 2005). This would apply accordingly to passengers suffering from flight phobia; that is, secondary issues such as claustrophobia (fear of confined spaces) or other psychiatric disorders may be involved as well.

6.10.2 SYMPTOMS

Symptoms of fear of flying are the same as those for other forms of anxiety and may include palpitations, dizziness, chest pressure, paleness, cold sweat, and a strong need to use the bathroom. Physical symptoms are often accompanied by feelings of surreality, feeling faint, or thinking that one is about to go insane. Characteristics of phobias normally include high levels of anxiety in everyday situations, an absence of rational explanations to the response, a loss of control over the response, and insufficient coping strategies to deal with the response.

A number of methods have been created to measure fear of flying. Examples are the flight anxiety modality (FAM) questionnaire and the flight anxiety situations (FAS) questionnaire (Van Gerwen al. 1999). In both questionnaires, the responders are asked to rate each item on a five-point Likert scale. Some questions (from FAM) relate to physical symptoms (1–3) and the second category is concerned with thought processes (4–6), as follows:

1. I am short of breath.
2. I feel dizzy, or I have the feeling that I'm going to faint.
3. I have the feeling that I'm going to choke.
4. I think the particular plane I am on will crash.
5. I attend to every sound or movement of the plane and wonder if everything is fine.
6. I continuously pay attention to the faces and behavior of the cabin crew.

Such measuring instruments can be used to establish the extent to which someone is afraid of flying and may also be used in research to study the effectiveness of treatments. It is also possible to employ physiological measurements (e.g., of the subject's heart rate) to complement the self-report.

6.10.3 WHAT IS THERE TO BE AFRAID OF?

One may ask what someone suffering from flight phobia worries about. Normally, affected individuals are concerned about conditions outside their control, such as poor weather conditions, technical failures, and human error (in pilots or flight controllers). In other cases, the source of worry is self-consciousness about one's own reactions and unusual behavior, such as fainting in the aircraft. Some find the whole experience leading up to a flight to be worse than the flight itself, including issues relating to check-in, queues, delays, and time-consuming security checks. On the other hand, air travel may enable positive experiences, although this aspect has been studied to a lesser extent than the negative implications of air travel.

6.10.4 TREATMENT

The chosen course of treatment depends on whether fear of flying is the only problem or if the person has other phobias, depression, or other mental illness. Studies show that people suffering from personality disorders benefit from standard treatments of flight anxiety (Van Gerwen et al. 2003). Thus, those suffering from mental illnesses should not avoid treatment for aviophobia; however, these individuals are unlikely to receive the full treatment required for separate disorders. A summary of treatment types revealed that most involved some form of mapping the individual's problem (Van Gerwen et al. 2004). Typically, treatment is of relatively short duration (1–3 days) and takes place in a group setting; the content usually consists of relaxation exercises, as well as some form of cognitive–behavioral therapy. In most cases, therapy concludes with exposure to flying in flight simulators or real-world aircraft. Recently, a number of studies have emerged that use personal computers as a medium for exposure to flying. These studies involve simulating a flight using visual, auditory, and physical stimuli (vibrations) combined with relaxation techniques, provisioning a framework to structuring emotions and stopping undesired trains of thought.

When asked which methods they use to help them relax, about 37% of passengers respond that they sometimes or often use alcohol to reduce anxiety related to flying. A total of 47% seek distractions and just below 10% use some form of medication. Booklets and information have been developed to advise passengers suffering from fear of flying how to cope with and prepare for a flight. These points are based on advice produced at a conference (Airborne, Vienna, December 2000) at which a number of flight phobia experts convened to discuss the topic (quoted in Van Gerwen et al. 2004). A number of organizations also promote the study and treatment of flight anxiety (such as The Valk Foundation, accessible at http://www.valk.org).

Advice to sufferers of fear of flying from Van Gerwen et al. (2004, p. 33) includes:

* Avoid caffeine, sugar, nicotine, and self-medication.
* Practice relaxation.
* Drink plenty of water and avoid alcohol. Alcohol does not decrease but rather increases fear and contributes to dehydration.
* Pay attention to your breathing and regularly carry out your breathing exercises.
* Turbulence is uncomfortable, but safe when your seatbelt is fastened.
* Stop the "what ifs" and focus on "what is."
* Keep flying. Do not avoid it.
* Motivation is the key to change.
* Planes are designed and built to fly.
* Write on cards reminders of personal coping instructions that work for you.

6.11   THE PAINS AND PLEASURES OF AIR TRAVEL

After the September 11, 2001, attacks on the United States, many people avoided air travel for a while, causing passenger figures to plummet. The number of travelers had more or less rebounded by 2003; however, the number of Americans traveling abroad was still in decline (Swanson and McIntosh 2006). In a British survey, a large majority of respondents (85%) said the September 11 events did not influence their future travel plans (Gauld et al. 2003). Such terrorist attacks are probably regarded as isolated events, causing the majority of the population to think they will not happen to them.

For some, however, this will be a source of concern, especially when combined with media attention on other health risks such as deep vein thrombosis and cardiovascular and infectious diseases, such as the recent emergence of severe acute respiratory syndrome (SARS). The flight-related increase in the occurrence of blood clots, for example, is extremely low except for individuals belonging to a particularly susceptible group (Bendz 2002; Owe 1998); however, there is often a difference between the real, objective risk that something bad will happen and the subjectively observed risk. Significant differences also exist between people in terms of risk aversion. Typically, people are willing to take greater risks when in a position of control or influence, which is rarely the case in air travel unless the person is piloting the aircraft.

In addition, everyday events or factors may act as stressors—for example, the purpose of the trip itself, such as doing something one does not look forward to or leaving one's family for a long period of time. Other aspects include the trip to the airport, check-in, and security controls, of which the conditions are constantly changing and becoming more stringent. It may be difficult to estimate the amount of time the boarding process will take and to decide whether delays will ensue. Sometimes many different things will go wrong at the same time, causing the accumulation of various stressors. People also differ in relation to personality and coping techniques and will therefore react in different ways to identical events, much in the same way that people react differently to work-related stress. In a British study, travelers were asked which factors led to anxiety, and the most common responses were delays (mentioned by 50% of respondents) and boarding the plane (mentioned by 42% of respondents) (McIntosh et al. 1998).

In a more recent study, Norwegian passengers were asked about the positives and negatives of air travel (Martinussen et al. 2008). A total of 268 individuals, consisting in part of travelers and in part of individuals recruited at the University of Tromsø, participated in the survey. Participants were asked how they usually experienced check-in and security controls. Of the respondents, 24% said they experienced check-in as either moderately or very stressful. In total, 11% of respondents reported dissatisfaction with the way check-in was conducted. With regard to security controls, 24% were dissatisfied with the way that they were conducted and a total of 40% reported experiencing this aspect of air travel as moderately or very stressful. Those who felt unable to influence events and felt that information was lacking at the airport were likely to be more stressed during check-in and when going though security; these respondents also experienced lesser degrees of pleasure in relation to air travel, as well as greater degrees of anxiousness (Martinussen et al. 2008).

Participants were asked about the positive sides of air travel; 72% said they looked forward to it, and 56% reported excitement about being in an airport. In other words, most people see positive sides of air travel, although there are negative aspects as well.

Swanson and McIntosh (2006) have launched a model for stress related to air travel. The model describes various factors that may be considered demands or stressors to which passengers are exposed. These factors involve check-in, security controls, the flight itself, and conditions in the cabin (e.g., poor air quality and limited space). The moderating variables are described as personality traits, coping mechanisms, and demographic variables such as gender. These variables may enhance or reduce the effects of stressors. The model also describes possible aftereffects of flight-related stress, such as physiological reactions, health complaints, aviophobia, and even anger and frustration. This model is similar to general stress models with the notable exception that it is not yet supported by much research, although the model is highly persuasive. The model can be extended to include resources that are thought to alleviate stress—for example, access to information and influence (or control) over the situation.

6.12   UNRULY PASSENGER BEHAVIOR

At 10,000 feet, a female passenger attempted to open the rear exit door of an aircraft traveling from Zurich, Switzerland, to Copenhagen, Denmark. Presumably, the woman was mentally ill and wanted to take her life. The cabin crew managed to overpower her, and she was handed over to the Copenhagen police force upon landing (*Dagbladet,* November 23, 2006). Another episode reported in the media involved a Norwegian male on a flight from Bangkok to Copenhagen who started to act in a rowdy manner (*Dagbladet,* November 13, 2003). The man was under the influence of alcohol and wanted to leave the aircraft. He was particularly loud, aggressive, and uncooperative. Several passengers got involved in the struggle to calm the man down, and a doctor on board gave him a sedative injection, which seemed to have little effect. The unruly passenger was finally brought under control and was strapped to his chair for the remainder of the flight, after which the Copenhagen police dealt with the ordeal.

These are but a few examples of situations in which aircraft passengers display behavior that violates the rules and regulations of air travel. Presumably, the media only cover more serious events, leaving minor incidents unreported. Examples of the latter typically include verbal abuse such as uttering insulting, harassing, or sexually charged statements to cabin crew members or fellow passengers.

Why did the woman want to leave the aircraft? Was she trying to take her life, or was she simply unaware of her surroundings? Similarly, one may wonder what provoked the episode involving the heavily intoxicated male on the flight from Bangkok and how alcohol affected the situation. In these types of scenarios, the consequences for the cabin crew and other passengers getting involved in or observing the incident must be considered. We will return to these questions after taking a closer look at the term "air rage."

6.12.1 What Is Air Rage?

Aggressive passengers or customers are not unique to aviation; they may be regarded generally as work-related violence, which occurs in many professions, including health care services, education, and service industries. This chapter will only investigate work-related violence initiated by passengers (not by co-workers). The motivating factors to such incidents are often real or imaginary inadequacies in the service provided or failure to meet demands, such as issues with seat reservations or the provision of additional alcohol. In some cases, the passenger may object to given instructions or refuse to accept current regulations pertaining to smoking or seat belts. The unique attribute of aviation is that a disorderly passenger cannot simply be "set off at the next stop" and that it is rather difficult to obtain additional assistance: As soon as the passengers have boarded and the aircraft has taken off, the crew is left to handle problems on its own.

Generally, a distinction is made between the terms *air rage* and *unruly behavior* or *misconduct*. An unruly passenger normally refuses to accept regulations that apply aboard the aircraft. He or she may engage in threats, abusive language, and/or noisy and inappropriate behavior, and refuse to follow instructions given by crew members. However, a disorderly passenger remains nonviolent. The notion of air rage applies to cases that involve physical violence. The problem of disorderly passengers and air rage occurs worldwide and, in the worst-case scenario, may be considered a safety threat. NASA investigations revealed that errors occurred in 15 of 152 cases in which one of the pilots had to get involved to subdue a disorderly passenger or was sought by cabin crew to help. Such errors included flying at the wrong altitude or choosing the wrong runway—potentially dangerous situations.

In addition to being a safety threat, air rage and unruly passengers often cause discomfort for the other passengers, cabin crew, and, potentially, the pilots. Additional landings and delays may be necessary. In some cases, the situation may escalate from dealing with a disorderly passenger to dealing with air rage (i.e., the person ends up resorting to violence). In other cases, there can be few indications that something is wrong, and a seemingly unprovoked attack on the crew may occur. An example of the latter was the case of a Norwegian Kato Air flight in 2004 from Narvik to Bodø (two cities in northern Norway) in which a passenger attacked the pilots with an ax, and a full-blown disaster was only marginally avoided. The passenger was later sentenced to 15 years' imprisonment. Such events may appear similar to terrorism; however, disorderly passengers and air rage are not considered to be politically or ideologically motivated.

6.12.2 How Frequently Does Air Rage Occur?

It is difficult to judge the extent of the problem of air rage and whether or not it is on the rise, mainly because its registration has been insufficient in many countries. The phenomenon first gained widespread attention in the mid-1990s. Whether people started taking notice of the problem because occurrence rates were rising or because changes in the aviation industry were causing dissatisfaction in customers, leading to a greater number of disorderly passengers, is a complex determination. Currently, little empirical evidence is available in this particular field. A possible explanation for this vacuum may be that it is difficult to define what constitutes unruly behavior (or air rage) and that the registration methods used by various airlines have not gone through the necessary standardization processes.

An analysis of the available statistics revealed disparities between different airlines, although several reported an increase in air rage and disorderly behavior (Bor 1999). British Airways reported 266 occurrences during a year (1997–1998) in which a total of 41 million passengers were transported. The probability of witnessing or experiencing air rage thus seems quite small. However, the reporting of these events may contain inadequacies; for example, it is fair to assume that only more serious events were reported.

6.12.3 What Causes Air Rage and Unruly Behavior?

A number of factors are said to contribute to, or be associated with, disorderliness in passengers; however, data to support these claims are lacking, and conclusions are often based on information from secondary sources (such as the media). A study based on several hundred media reports on the issue found that, essentially, three factors were associated with disorderly or violent passengers: alcohol, nicotine deprivation, and mental illness (Anglin et al. 2003). These are all factors that can be amplified by environmental stressors, such as confined spaces, poor air quality, delays, and, perhaps, poor service. A typical example of air rage involves a male passenger with a history of violent behavior who chain smokes and consumes a large amount of alcohol daily. He continues drinking on board and is infuriated when refused the opportunity to smoke. The study estimated that alcohol was a contributing factor in 40% of cases (Anglin et al. 2003).

Other studies based on reports from companies in the United States and Britain have found that alcohol was a contributing factor in 43% and 50% of the reported cases, respectively (Connell, Mellone, and Morrison 1999). In nearly every other incident, the violent passenger had started drinking before boarding the flight. Although we have limited knowledge about how alcohol interacts with other factors, we can assume that it is unlikely to be the only factor. It is more likely that alcohol reduces the person's inhibitions and sense of judgment; in a high-stress situation, this leads him or her to use aggressive behavior as a solution to problems.

Generally, few data are available to describe typical characteristics of disorderly passengers. They may be traveling alone, a couple, or a group traveling together. Sometimes socioeconomically well-placed individuals, such as artists or lawyers, have been involved in incidents using verbal and physical abuse. Perhaps such personality types are unaccustomed to taking orders or guidance from other people; they may not accept instructions from someone they perceive as of a lower rank than themselves. The cabin crew, on the other hand, must alternate between servicing passengers and being responsible for on-board safety. To some passengers, this duality may be difficult to accept because they expect service rather than instructions or commands.

The phenomenon of air rage and disorderly passengers may be better understood in light of general stress theory, with the addition of a number of contributing factors.

The cause of disorderliness, then, lies in the exposure to a sequence of events, each of which acts as a source of frustration (delays, queuing, and security checks). In addition, this is combined with alcohol and a triggering event such as not being allowed to take carry-on luggage on board. Moreover, many people are anxious about flying to some extent, which may be a factor as well. In some cases, the cabin crew's behavior and communication with the disorderly individual can contribute to the incident developing in a negative direction. For example, raising one's voice to the individual may be necessary to make oneself heard in the cabin because there is usually considerable background noise; however, the upset individual may perceive the behavior as reproving and impolite.

6.12.4   What Can Be Done to Prevent Air Rage?

Airlines provide cabin crew with training in dealing with disorderly passengers, including worst-case scenarios that involve physically strapping passengers to their seats. Some companies practice zero-tolerance policies in the event of air rage, resulting in police prosecution and denial of future flights with the company. Another important factor is to train cabin crew in how to manage frustrated and upset passengers in general. Naturally, crew members are required to respond to complaints and expressions of inconvenience in a polite and respectful manner. Providing information about delays and other irregularities is also important. Part of the training involves recognizing and preempting problems caused by passengers suffering from mental illness or under the influence of alcohol. Crucial preventive measures include refusing intoxicated persons permission to board and limiting access to alcohol in airports and, not least, on board. Training may also consist of learning how to calm a person down in a polite manner without causing the situation to escalate.

6.13   SUMMARY

This chapter has discussed individual differences and common ways in how people react to different stressors in life. We have established that when the experienced level of stress exceeds the amount with which a person is able to cope, various emotional, cognitive, and physiological reactions emerge. These reactions are of significance to one's general health condition, work achievements, performance, and job satisfaction. Stress has both short-term and long-term effects on the individual, and it is important to be familiar with these effects for one's own sake and because most aviation professions demand significant cooperation with colleagues and others. The chapter has mostly related to persons working in aviation, although passenger issues have been described to a certain extent.

RECOMMENDED READING

Bor, R., and Hubbard, T. 2006. *Aviation mental health: Psychological implications for air transportation*. Aldershot, England: Ashgate.
Dahlberg, A. 2001. *Air rage. The underestimated safety risk*. Aldershot, England: Ashgate.

REFERENCES

Åkerstedt, T. 2007. Altered sleep/wake patterns and mental performance. *Physiology and Behavior* 90:209–218.

Allen, T. D., Herst, D. E. L., Bruck, C. S., and Sutton, M. 2000. Consequences associated with work-to-family conflict: A review and agenda for future research. *Journal of Occupational Health Psychology* 5:278–308.

Anglin, L., Neves, P., Giesbrecht, N., and Kobus-Matthews, M. 2003. Alcohol-related air rage: From damage control to primary prevention. *Journal of Primary Prevention* 23:283–297.

Bendz, B. 2002. Flyreiser og venøs trombose [Air travel and deep vein thrombosis]. *Tidsskrift for Den Norske Legeforening* [*Journal of Norwegian Medical Association*] 122:1579–1581.

Block, J. 1995. A contrarian view of the five-factor approach to personality description. *Psychological Bulletin* 117:187–215.

Bor, R. 1999. Unruly passenger behaviour and in-flight violence. A psychological perspective. In *Proceedings of the Tenth International Symposium on Aviation Psychology*, ed. R. Jensen, B. Cox, J. Callister, and R. Lavis. Columbus: Ohio State University.

Bourgeois-Bougrine, S., Carbon, P., Gounelle, C., Mollard, R., and Coblentz, A. 2003. Perceived fatigue for short and long-haul flights: A survey of 739 airline pilots. *Aviation, Space, and Environmental Medicine* 74:1072–1077.

Bradley, R., Greene, J., Russ, E., Dutra, L., and Westen, D. 2005. A multidimensional meta-analysis for psychotherapy for PTSD. *American Journal of Psychiatry* 162:214–227.

Clarke, S., and Robertson, I. T. 2005. A meta-analytic review of the big five personality factors and accident involvement in occupational and nonoccupational settings. *Journal of Occupational and Organizational Psychology* 78:355–376.

Connell, L., Mellone, V. J., and Morrison, R. M. 1999. Cabin crew safety information article. In *Proceedings of the Tenth International Symposium on Aviation Psychology*, ed. R. Jensen, B. Cox, J. Callister, and R. Lavis. Columbus: Ohio State University.

Costa, G. 1996. The impact of shift and night work on health. *Applied Ergonomics* 27:9–16.

———. 2003. Factors influencing health of workers and tolerance to shift work. *Theoretical Issues in Ergonomics Science* 4:263–288.

Costa, G., Sartori, S., and Åkerstedt, T. 2006. Influence of flexibility and variability of working hours on health and well-being. *Chronobiology International* 23:1125–1137.

Costa, P. T., and McCrae, R. R. 1985. *The NEO personality inventory manual.* Odessa, FL: Psychological Assessment Resources.

———. 1997. Personality trait structure as a human universal. *American Psychologist* 52:509–516.

Demerouti, E., Geurts, S. A., Bakker, A., and Euwema, M. 2004. The impact of shiftwork on work–home conflict, job attitudes and health. *Ergonomics* 47:987–1002.

Digman, J. M. 1990. Personality structure: Emergence of the five-factor model. *Annual Review of Psychology* 41:417–440.

Ekeberg, Ø., Seeberg, I., and Ellertsen, B. B. 1989. The prevalence of flight anxiety in Norway. *Nordic Journal of Psychiatry* 43:443–448.

Folkard, S., and Tucker, P. 2003. Shift work, safety, and productivity. *Occupational Medicine* 53:95–101.

Frankenhaeuser, M. 1991. The psychophysiology of sex differences as related to occupational status. In *Women, work, and health. Stress and opportunities,* ed. M. Frankenaeuser, U. Lundberg, and M. Chesney, 39–61. New York: Plenum Press.

Friborg, O., Barlaug, D., Martinussen, M., Rosenvinge, J. H., and Hjemdal, O. 2005. Resilience in relation to personality and intelligence. *International Journal of Methods in Psychiatric Research* 14:29–42.

Gauld, J., Hirst, M., McIntosh, I. B., and Swanson, V. 2003. Attitudes to air travel after terrorist events. *British Travel Health Association Journal* 3:62–67.

Goldberg, L. R. 1993. The structure of phenotypic personality traits. *American Psychologist* 48:26–34.

Karasek, R. A. 1979. Job demands, job decision latitude, and mental strain: Implications for job redesign. *Administrative Science Quarterly* 24:285–308.

Karasek, R. A., and Theorell, T. 1990. *Healthy work: Stress, productivity, and the reconstruction of working life.* New York: Basic Books.

Kitterød, R. H. 2005. *Han jobber, hun jobber, de jobber. Arbeidstid blant par av småbarnsforeldre* [*He works, she works, they work. Work hours among couples with small children*]. Oslo: Statistics Norway.

Klein, D. E., Brüner, H., and Holtman, H. 1970. Circadian rhythm of pilot's efficiency, and effects of multiple time zone travel. *Aerospace Medicine* 41:125–132.

Klein, G. A. 1995. A recognition-primed decision making (RPD) model of rapid decision making. In *Desicion models in action: Models and methods,* ed. G. A. Klein, J. Orasanu, R. Calderwook, and C. E. Zsambok, 138–147. Norwood, NJ: Ablex Publishing Corporation.

Knutsson, A. 2003. Health disorders of shift workers. *Occupational Medicine* 53:103–108.

Lazarus, R. S., and Folkman, S. 1984. *Stress, appraisal and coping.* New York: Springer.

Leiter, M., and Maslach, C. 2005. *Banishing burnout. Six strategies for improving your relationship with work.* San Francisco, CA: Jossey–Bass.

Lundberg, U. 2005. Stress hormones in health and illness: The roles of work and gender. *Psychoneuroendocrinology* 30:1017–1021.

Lundberg. U., and Frankenhaeuser, M. 1999. Stress and workload of men and women in high-ranking positions. *Journal of Occupational Health Psychology* 4:142–151.

Lundberg, U., Mårdberg, B., and Frankenhaeuser, M. 1994. The total workload of male and female white collar workers as related to age, occupational level, and number of children. *Scandinavian Journal of Psychology* 35:315–327.

Martinussen, M., and Richardsen, A. M. 2006. Job demands, job resources, and burnout among air traffic controllers. *Aviation, Space, and Environmental Medicine* 77:422–428.

Martinussen, M., Richardsen, A. M., and Burke, R. J. 2007. Job demands, job resources and burnout among police officers. *Journal of Criminal Justice* 35:239–249.

Martinussen, M., Gundersen, E., and Pedersen, R. 2008. The joys and stressors of air travel. Paper presented at the 28th EAAP conference in Valencia, Spain, Oct. 27–31.

Maslach, C., and Jackson, S. E. 1981. The measurement of experienced burnout. *Journal of Occupational Behavior* 2:99–113.

———. 1986. *Maslach burnout inventory manual,* 2nd ed. Palo Alto, CA: Consulting Psychologists Press, Inc.

Maslach, C., Schaufeli, W., and Leiter, M. P. 2001. Job burnout. *Annual Review of Psychology* 52:397–422.

McIntosh, I. B., Swanson, V., Power, K. G., Raeside, F., and Dempster, C. 1998. Anxiety and health problems related to air travel. *Journal of Travel Medicine* 5:198–204.

Medialdea, J., and Tejada, F. R. 2005. Phobic fear of flying in aircrews: Epidemiological aspects and comorbidity. *Aviation, Space, and Environmental Medicine* 76:566–568.

Megdal, S. P., Kroenke, C. H., Laden, F., Pukkala. E., and Schernhammer, E. S. 2005. Night work and breast cancer risk: A systematic review and meta-analysis. *European Journal of Cancer* 41:2023–2032.

Ng, T. W. H., Sorensen, K. L., and Eby, L. T. 2006. Locus of control at work: A meta-analysis. *Journal of Organizational Behavior* 27:1057–1087.

Nicholson, A. N. 2006. Sleep and intercontinental flights. *Travel Medicine and Infectious Disease* 4:336–339.

Orasanu, J. 1997. Stress and naturalistic decision making: Strengthening the weak links. In *Decision making under stress. Emerging themes and applications,* ed. R. Flin, E. Salas, M. Strub, and L. Martin, 43–66. Aldershot, England: Ashgate.

Orth, U., and Wieland, E. 2006. Anger, hostility, and posttraumatic stress disorder in trauma-exposed adults: A meta-analysis. *Journal of Consulting and Clinical Psychology* 74:698–706.

Østlyngen, A., Storjord, T., Stellander, B., and Martinussen, M. 2003. En undersøkelse av total arbeidsbelastning og tilfredshet for psykologer i Norge [A survey of total work-load and work satisfaction among psychologists in Norway]. *Tidsskrift for Norsk Psykologforening* [*Journal of the Norwegian Psychological Association*] 40:570–581.

Owe, J. O. 1998. Helsemessige problemer hos flypassasjerer [Health problems in airline passengers]. *Tidsskrift for Den Norske Legeforening* [*Journal of the Norwegian Medical Association*] 118:3623–3627.

Ozer, E. J., Best, S. R., Lipsey, T. L., and Weiss, D. S. 2003. Predictors of posttraumatic stress disorder and symptoms in adults: A meta-analysis. *Psychological Bulletin* 129:52–73.

Pallesen, S. 2006. Søvn [Sleep]. In *Operativ psykologi* [*Operational psychology*], ed. J. Eid and B. J. Johnsen, 196–215. Bergen, Norway: Fagbokforlaget.

Richardsen, A. M., and Martinussen, M. 2006. Måling av utbrenthet: Maslach burn-out inventory [Measuring burnout: The Maslach burnout inventory]. *Tidsskrift for Norsk Psykologforening* [*Journal of the Norwegian Psychological Association*] 43:1179–1181.

Rosenman, R. H., and Friedman, M. 1974. Neurogenic factors in pathogenesis of coronary heart disease. *Medical Clinics of North America* 58:269–279.

Schaufeli, W. B., and Bakker, A. 2004. Job demands, job resources, and their relationship with burnout and engagement: A multisample study. *Journal of Organizational Psychology* 25:293–315.

Schultz, M. S., Cowan, P. A., Pape Cowan, C., and Brennan, R. T. 2004. Coming home upset: Gender, marital satisfaction, and the daily spillover of workplace experience into couple interactions. *Journal of Family Psychology* 18:250–263.

Selye, H. 1978. *The stress of life.* New York: McGraw–Hill.

Silvera, D., Martinussen, M., and Dahl, T. I. 2001. The Tromsø social intelligence scale, a self-report measure of social intelligence. *Scandinavian Journal of Psychology* 42:313–319.

Swanson, V., and McIntosh, I. B. 2006. Psychological stress and air travel: An overview of psychological stress affecting airline passengers. In *Aviation mental health: Psychological implications for air transportation,* ed. R. Bor and T. Hubbard, 13–26. Aldershot, England: Ashgate.

Terracciano, A., Costa, P. T., and McCrae, R. R. 2006. Personality plasticity after age 30. *Personality and Psychology Bulletin* 32:999–1009.

Tüchsen, F., Hannerz, H., and Burr, H. 2007. A 12 year prospective study of circulatory disease among Danish shift workers. *Occupational Environmental Medicine* 63:451–455.

Ursin, R. 1996. *Søvn: En lærebok om søvnfysiologi og søvnsykdommer* [*Sleep: A textbook on sleep physiology and sleep pathology*]. Oslo: Cappelen Akademisk Forlag.

Van Gerwen, L. J., Spinhoven, P., Van Dyck, R., and Diekstra, R. F. W. 1999. Construction and psychometric characteristics of two self-report questionnaires for the assessment of fear of flying. *Psychological Assessment* 11:146–158.

Van Gerwen, L. J., Delorme, C., Van Dyck, R., and Spinhoven, P. 2003. Personality pathology and cognitive–behavioral treatment of fear of flying. *Journal of Behavior Therapy and Experimental Psychiatry* 34:171–189.

Van Gerwen, L. J., Diekstra, R. F. W., Arondeus, J. M., and Wolfger, R. 2004. Fear of flying treatment programs for passengers: An international update. *Travel Medicine and Infectious Disease* 2:27–35.

Waage, S., Pallesen, S., and Bjorvatn, B. 2006. Skiftarbeid og søvn [Shift work and sleep]. *Tidsskrift for Norsk Psykologforening* [*Journal of the Norwegian Psychological Association*] 44:428–433.

Westman, M., and Etzion, D. 1995. Crossover of stress, strain and resources from one spouse to another. *Journal of Organizational Behavior* 16:169–181.

Yoshimasu, K. 2001. Relation of type A behavior pattern and job-related psychosocial factors to nonfatal myocardial infarction: A case-control study of Japanese male workers and women. *Psychosomatic Medicine* 63:797–804.

7 Culture, Organizations, and Leadership

7.1 INTRODUCTION

This chapter discusses organizational and cultural factors and how these factors influence people working in aviation. The aviation industry is an international business in which individuals with different cultural backgrounds must work together to make sure that aircraft arrive at their destination in a safe and timely manner. Communication problems can lead to irritation and disagreement and may even have serious safety repercussions. Communication and coordination are always demanding, especially when people have different cultural backgrounds, genders, and languages. Toward the end of this chapter, we discuss organizational changes and leadership and how these influence employees and the jobs that need to be done.

7.2 DO ORGANIZATIONAL ISSUES PLAY A ROLE IN ACCIDENTS?

In the past decades, a number of significant accidents have occurred, such as the 1986 meltdown at the nuclear power station in Chernobyl, the explosion at the North Sea Piper Alpha oil rig in 1988, and, more recently, the space shuttle *Columbia* disaster (2003). These accidents have in common that organizational factors were mentioned as contributing causes (Pidgeon and O'Leary 2000). The constructs of culture or safety culture are often mentioned as part of an explanation of what causes accidents or problems within the organization.

In several preceding chapters, we have focused on individuals and individual differences. This has included selection and training as well as their importance to a person's performance. In this chapter, we will look at organizational issues and aspects of the system that may be significant for how a person acts and, not least, the consequences of these actions for safety.

Major accidents, whether they are plane crashes or nuclear disasters, have tremendous human, economic, and environmental consequences. Thus, attempting to unveil the causes of accidents is not something new. Wiegmann and colleagues (2004) describe various historical stages in how attempts have been made to explain such accidents. The first stage (technical period) was distinguished by rapid technological development, during which investigators looked for inadequacies and outright flaws in technical systems. In the next phase, focus shifted toward human error (period of human error), and investigators sought to find errors in the human operator. This was followed by the sociotechnical stage, where the interface between human operators and technology was examined. In the final phase, which the authors labeled the organizational culture period, individuals are no longer regarded as operators of machinery isolated from the world at large, but rather as workers in a team within a given cultural context.

7.3   WHAT IS CULTURE?

The term "culture" has been associated with organizations since the early 1980s. There are about as many definitions of culture as there are publications about it. By "organizational culture," one normally means, somewhat inaccurately, "the way things are done around here." A more formal definition can be found in the book on organizational culture by Henning Bang (1995, p. 23): "Organizational culture is the set of commonly shared norms, values, and perceived realities which evolve in an organization as its members interact with each other and their surroundings." Schein (1996, p. 236) defines the term as the "set of shared, taken for granted implicit assumptions that a group holds, and that determines how it perceives, thinks about, and reacts to various environments." These definitions differ in the sense that the former describes how the culture takes shape (i.e., through interaction), whereas the latter describes its effect on the members of the group.

There are also a number of other, related terms, such as *climate*. Many articles and empirical studies use these terms interchangeably—that is, without clarifying their differences. Thus, scientists studying these phenomena are mapping approximately the same thing, although they appear to use different constructs for what these things are (Mearns and Flin, 1999, provide a summary).

Schein (1990) considers climate to be a manifestation and measurable aspect of culture. Thus, culture is a deeper phenomenon that is not easily charted or categorized. Others say that although culture is what is shared, or common, for members, climate is a kind of "average" of the group members' experience—preferably the interpersonal relationships within the organization. The last word has hardly been spoken on this matter. In part, it is likely that the constructs have different histories and associated measurement techniques; however, the subjects of these studies are presumably overlapping phenomena. Climate is usually measured using standardized scales—an approach critics label as insufficient to get hold of the culture (see, for example, Schein 1990). Methodologies, including interviews and observation, are, then, the preferred alternatives. Most empirical studies of culture, however, have used questionnaires to measure the construct.

As it relates to definitions of culture, the term "organization" applies to both businesses (e.g., airlines) and groups composed in different ways, such as pilots (a profession) or subgroups within a business (e.g., women or technicians). It is common to study national cultures—that is, the extent to which differences exist between nations. In aviation, therefore, we may assume that individuals are affected by several cultures: national, professional, and company (the airline for which the person works) cultures. Cultures may develop in many different social systems as people interact over time. According to Schein (1990), the conditions necessary for cultural development include that the individuals must have worked together for a long enough time to have experienced and shared important problems. They must have had the opportunity to solve these problems and observe the effect of implemented solutions. Last, but not least, the group or organization must have taken in new members who have been socialized into the way the group thinks, feels, and solves problems. The advantage of having a culture is that it makes events more predictable and gives things meaning, which may reduce anxiety in group members (Schein 1990).

Subcultures, which may be in conflict or support each other, can also form within an organization. They may be based on profession, workplace (sea versus land), gender, or age. In the wake of corporate mergers, subcultures can form based on the formerly separate companies. In conflicts between subgroups, each side will typically view the other from a polarized, black-and-white perspective: "They are bad. We are good—we have the correct values." One explanation for the rise of such conflicts may be that groups have a need to preserve their social identity and will defend themselves against those who want to destroy or threaten their culture (Bang 1995). However, when such conflicts are allowed to thrive, they can sometimes be devastating for an organization, with harmful consequences for the well-being and health of individuals in the worst-case scenario. Subcultures arise in most organizations, and it is probably naïve to think that conflicts will never arise between them. Presumably, how the organization and its leadership manage such conflicts would be more important than preventing their rise.

7.4   NATIONAL CULTURE

Aviation is, in almost every respect, an international industry. This forces companies and individuals to interact with people from other cultures, who often have a language other than English as their first language. National culture affects the way people communicate and act. The most popular model and method for studying these national differences are based on Geert Hofstede's questionnaire for work-related values (Hofstede 1980, 2001). Hofstede developed the measurement instrument in connection with an extensive study of IBM employees in 66 countries conducted from 1967 to 1973. Participants noted the importance or significance of given values—that is, the extent to which they agreed or disagreed with the statements. The questions were placed in four scales (power-distance—PD; uncertainty-avoidance—UAV; individualism-collectivism—IND; and masculinity-femininity—MAS). Descriptions of these with corresponding sample statements are given in Table 7.1.

Other scientists have used this measurement instrument in corresponding cross-cultural studies, and a reasonable amount of support for these four dimensions has been established. However, the instrument has been criticized, notably for its low internal consistency as to the different scales (Spector and Cooper 2002; Hofstede 2002). Hofstede's study is nonetheless both impressive and important because of the great number of countries and individuals that participated and because it facilitates comparisons between his data and other findings.

A study by Merrit (2000) that included almost 10,000 pilots from 19 countries revealed that two of the four dimensions (IND and PD) were replicable, whereas there were issues with some of the original questions for masculinity and UAV. However, a clear correspondence was present between Hofstede's ranking of countries according to cultural dimensions and the scores of pilots from the various countries. In addition, there were some differences between pilots as a group and

TABLE 7.1
Hofstede's Scales for Measuring National Culture

	What Does It Measure?	Sample Question
Power-distance (PD)	Denotes the degree to which power is unequally distributed between managers and subordinates and the extent to which this is accepted. Low PD values have been recorded in Austria, Israel, and the Scandinavian countries; high PD levels have been found in the Philippines and Mexico.	How often are employees afraid to express disagreement with their managers?
Uncertainty-avoidance (UAV)	Denotes the extent to which members of a culture feel threatened or anxious due to uncertainty and unpredictable situations. Countries including Greece, Portugal, and several Latin American countries report high levels of UAV; the United States, Singapore, Sweden, and Denmark report low UAVs.	How often do you feel nervous and tense at work?
Individualism-collectivism (IND)	The extent to which focus is on the individual (i.e., the individual's rights and responsibilities versus the group's). High levels of individualism are found in Western nations such as the United States; several countries in Asia have low scores.	How important is it that you fully use your skills and abilities on the job?
Masculinity-femininity (MAS)	Measures the extent to which the culture emphasizes efficiency and competition versus more social (feminine) values. Countries with high femininity scores include the Scandinavian countries; Japan and some nations in Southern Europe and Latin America have low scores.	How important is it to have security of employment?

Source: Based on Hofstede, G. 1980. *Culture's consequences: International differences in work-related values.* Beverly Hills, CA: Sage.

the IBM employees used as a reference by Hofstede. For example, pilots had higher PD than Hofstede's group, underlining aspects of the piloting profession in which a clear-cut hierarchy (i.e., between captains and co-pilots) exists and is generally accepted.

A study by Sherman, Helmreich, and Merrit (1997) found differences between countries in terms of how pilots regard rules and procedures, the usefulness of automation, and the extent to which they accepted a definitive hierarchy (chain of command) between captains and co-pilots.

In another study of military pilots from 14 NATO countries, the values on Hofstede's scales were compared to accident statistics for those countries (Soeters and Boer 2000). Data were collected over a 5-year period (1988–1992), and the number of lost planes per 10,000 flight hours was used to describe the accident ratio. This number was then correlated with the national values for the four cultural dimensions. Three out of the four dimensions were significantly correlated with results for IND ($r = -.55$), PD ($r = .48$), and UAV ($r = .54$) (Soeters and Boer 2000). There was no significant correlation between the masculinity index and accident ratio. Correlations increased when accidents due to mechanical failure were removed. The numbers thus indicate greater accident rates in countries that have low individualism scores and high power-distance and uncertainty-avoidance scores.

The results are interesting; however, it is important to keep in mind that the number of countries involved was only $N = 14$ because the nations (not the pilots) were the subjects of the study. In other words, the correlations are based on a fairly small sample. In addition, a correlation is not the same as causality. Many other factors vary among countries, and these factors may cause the observed variations in the number of accidents. In addition, culture was not measured in the military pilots, but the results from Hofstede's study were used. Therefore, it is possible that figures are slightly different from what they would have been if culture had been measured in military pilots for the same time frame in which the accident statistics were collected. However, the study indicates that cultural factors and how they relate to accidents may be worth further investigation.

A third study (Li, Harris, and Chen 2007) compared accident statistics and accident causes from India, Taiwan, and the United States. A total of 523 accidents, including a combined 1,762 cases of human error, were investigated. The study summarized results from former accident surveys that used the HFACS system (Wiegmann and Shappell 2003) to classify errors. This system is based on Reason's model, which is discussed in greater detail in Chapter 8. The study found significant differences among countries in what were reported as causes. Organizational causes were reported more frequently in Taiwan and India (countries with high power-distance and low individualism scores) than in the United States. This suggests a hierarchy in which employees expect to be told what to do (to a greater extent than in Western nations) and collective decisions are preferred to individual decisions. The authors see this aspect as a possible explanation for more frequent reporting of organizational errors. There is less spontaneous feedback in the system by the means of open discussion, and subordinates have less authority and autonomy in decision making and, perhaps, in correcting errors and flaws.

7.4.1 PROBLEMS RELATING TO STUDY OF CULTURAL DIFFERENCES

Studying national differences in culture or, for that matter, other aspects is a complicated job. Completing the study or survey in approximately the same way in vastly different countries represents one of the challenges. For example, is it possible to sample participants in a similar way, and is the same procedure used in all the countries included in the study? The nature of the matter is that the more different the countries are, the more difficult it becomes to complete such a task. There may be different regulations as to the available registries of the target group and whether permission will be granted to extract information from these groups. For example, comparing nurses from The Netherlands to Malaysian pilots would hardly make a good basis for attributing potential findings to cultural differences alone. Ideally, the groups in question should be as similar as possible, even though researchers typically have to admit that, in practice, it is impossible to complete a survey in exactly the same way in all countries. Greater similarity between groups (in terms of other variables) generates a greater degree of certainty to conclusions that differences are due to cultural factors.

Another issue is represented by the challenge of translating questionnaires between different languages. Even though a substantial amount of effort is put into translations, a statement may convey a different meaning in another language, or certain words and expressions may not exist in the target language or correspond to the original one. Often, a translation is made from the original language to the target language (e.g., English to Norwegian). Then, someone else who is also proficient in both languages translates the text back into the original language (in this case, English). Finally, the two English versions are compared, and at this point ambiguities in the translators' efforts and what needs to be adjusted become clear. The most important point here is that a perfect, word-for-word translation is not necessarily desirable; the important thing is to preserve meaning. Finally, issues may arise in cross-cultural studies because people from different cultures have different response styles. This means that some cultures may have a greater tendency to agree to a specific statement; in other cultures, opinion is expressed more freely and the extreme ends of the scale are used to a greater extent.

7.5   PROFESSIONAL CULTURE

Many occupations or professions have strong cultural identities. This applies to psychologists, air traffic controllers, and pilots, to mention but a few examples. Often, there is fierce competition to be selected and successfully complete the required (and often extensive) education. Upon completion of training, many people within the profession join powerful unions, which act to preserve the rights and interests of members. Some unions offer members sponsored education to maintain or build additional skills; they also typically provide guidelines for ethical behavior within the profession. Thus, unions help to socialize new members into the group (profession) by exercising control over the members to some extent. Professionals, including pilots, psychologists, and physicians, are often highly enthusiastic and proud workers. They will make every effort to be successful, and few people quit the profession after having entered the work force. On the other hand, a strong professional culture can give individuals a false sense of invulnerability and disregard for their own limitations, according to Helmreich and Merrit (1998). Some people may be aware of various human limitations in general, but unaware of this applying to them (a form of unrealistic optimism).

A study by Helmreich and Merrit (1998, p. 35) revealed that a large proportion of pilots and physicians strongly believed that they were able to do their job just as well in below-average conditions—that is, that equally sound judgments were made in an emergency compared to under normal conditions or that personal issues were put on hold while working. There were also some differences between the groups; 60% of doctors declared they still worked efficiently when tired, compared to only 30% of pilots. Together, the doctors' perceptions were somewhat less realistic than the pilots'. What caused this distinction is not yet clear. However, the long-term focus on human factors in aviation may have made its impact. For example, pilots must complete mandatory CRM courses, whereas this is relatively new to medical professions.

Within a profession, there may be subgroups with which an individual feels more or less associated. Subgroups may be based on specialization, workplace, and/or gender. Examples include military versus civilian pilots or clinical psychologists versus psychology professors.

One study examined cultural changes in a major Norwegian airline, with a sample of 190 pilots (Mjøs 2002). Significant differences were found between the scores for this group and national norms based on Hofstede's figures. The greatest difference was found for the masculinity index, where pilots scored much higher than expected.

An international study of flight controllers from Singapore, New Zealand, and Canada investigated how they perceived their work environment (Shouksmith and Taylor 1997). The idea was that there would be greater differences between flight controllers from an Eastern culture compared to the two Western countries. The flight controllers were asked about what they considered stressful in their work environment, and many similarities were found. For example, flight controllers from all three countries mentioned technical limitations, periods of high traffic, and fear of causing accidents among the top five most important sources of stress.

On the other hand, Singaporean flight controllers also mentioned problems with local management in the top five, whereas the two remaining groups mentioned the general working environment as a top five stressor. The authors attribute this difference partly to cultural factors such as higher power-distance in Asia compared to Western nations, causing more severe implications when disagreements arise between subordinates and management. Even external environmental factors may explain some of the differences; for example, more frequently occurring bad weather in Canada could have led flight controllers to mention this factor as one of the most significant work-related stressors.

7.6 ORGANIZATIONAL CULTURE

As in national and professional cultures, there may be cultural variations between airlines operating in a country. In a Norwegian study including three airlines (Mjøs 2004), significant differences were found in three of the four Hofstede dimensions (PD, MAS, and UAV). The participating pilots ($N = 440$) were also asked to report any errors made during the past year as part of the survey. A total of 10 indicators, such as forgetting important checklist points or choosing the wrong taxi runway, were included. For each pilot, a total error score was calculated and correlated with the stated cultural dimensions (PD, UAV, IND, and MAS). A strong association ($r = .54$) was found between PD and the number of operational errors (i.e., the occurrence of errors increased with perceived PD). As mentioned earlier, correlations are not necessarily evidence of cause and effect; however, the result is interesting. We may only speculate as to mechanisms behind such correlations. Kjell Mjøs (2004) suggests a model where cultural variables have consequences for the social environment in a cockpit, which has consequences for communication between pilots that, in turn, can lead to operational errors.

Ten years after the first data were collected, a follow-up survey was conducted with a smaller selection of pilots from the largest airline from the previous study (Mjøs 2002). The purpose of this study was to investigate potential changes in the airline's culture over time. Mjøs found a significant change in the scores for dimensions PD, IND, and UAV. The social climate had also improved in this period. Some pilots were evaluated while performing an exercise in a simulator. A lower number of operational errors were recorded in the follow-up study compared to 10 years earlier. One of the study's weaknesses was that it did not survey exactly the same persons both times; however, this would obviously be difficult to accomplish due to the gap in time between the two surveys. Additionally, because the surveys were anonymous, it would have been impossible to record individual scores over time. It is also important to be aware that the study explains neither the cause of the cultural changes nor whether the recorded performance improvements can be attributed to these changes.

7.7 SAFETY CULTURE

Safety culture is a term that has been widely used in aviation and also, to some extent, in other industries where consequences of error are significant—for example, in high-tech factories and nuclear power plants, in surgery rooms, and in various modes of transport. Such systems often involve close interaction between technology and human operators, and errors may have disastrous consequences. There are a number of definitions of safety culture (for a summary, see Wiegmann et al. 2004). A relatively simple definition is that safety culture refers to the fundamental values, norms, presumptions, and expectations that a group shares concerning risk and safety (based on Mearns and Flin 1999).

Like organizational culture and organizational climate, boundaries between what is meant by safety culture and safety climate are unclear. Some people are of the opinion that they are different but related terms. That is, safety climate is measured using a questionnaire and provides a snapshot of how the employees perceive safety (often in relation to a specific issue), whereas safety culture refers to more lasting and fundamental values and norms that partially overlap the national culture of which the organization is a part (Mearns and Flin 1999). In practice, however, the notions are used interchangeably, and quantitative questionnaires often overlap in terms of content.

Wiegmann and colleagues (2004) have described a number of traits or presumptions that are shared in the different definitions of safety culture. These commonalities are that safety culture is something a group of people have in common, it is stable over time, and it is reflected in the organization's will to learn from mistakes, events, and accidents. Safety culture influences group members' behavior, either directly or through affecting the employees' attitudes and motivation to behave in a way that enhances safety.

7.7.1 What Characterizes a Sound Safety Culture?

An important aspect in a sound safety culture is management commitment and involvement in the promotion of safety. To achieve this, it is crucial that the highest levels of management make the necessary resources available and support the work involved. It must be reflected in all aspects of the organization, and routine evaluation and system improvements must take place. However, not only the higher levels of management but also lower level administrators, who should participate in activities related to improving safety, are important. Little is gained by sending employees to safety classes if those who are monitoring the implementation of routines do not participate.

Another indicator of safety culture is that those who are performing the specific jobs are given the responsibility and authority to be the last resort in case of errors. In other words, they feel enabled, and they regard their role as an important part in securing safety. This involves playing an active role and being heard in the work to improve safety. The organization's reward system is another aspect. Are reward systems in place to promote safety, or are employees punished or neglected when taking on an issue? A final concern is the extent to which the organization is willing to learn from previous mistakes and that employees are given feedback through the reporting system. Encouraging employees to report errors and mistakes but doing nothing to correct them would be very demotivating for those who do so.

7.7.2 How Does a Safety Culture Develop?

Safety cultures can be categorized in several ways, and they are often based on elements previously mentioned in the previous section. Hudson (2003) has developed a model for various safety cultures—from pathological to mature, or developed, cultures. The different categories are presented in Table 7.2 with sample statements for typical ways of regarding safety. This model expands on Westrum's model (Westrum and Adamski 1999), which contained three stages or organization types: *pathological, bureaucratic,* and *generative.*

TABLE 7.2
Development of a Safety Culture

	Type	Statement Typical of Culture
High-level information flow and trust	Generative	Safety is the way we do business around here.
	Proactive	We work on the problems that we still find.
	Calculative	We have systems in place to manage all hazards.
	Reactive	Safety is important. We do a lot every time we have an accident.
Low degree of flow of information and trust	Pathological	Who cares as long as we are not caught?

Source: Based on Hudson, p. 2003. *Quality and Safety in Health Care* 12:7–12.

The five stages in Hudson's model provide a framework for classifying safety cultures and describing various levels of cultural maturity. In *pathological* cultures, safety is seen as a problem caused by operators. Making money and avoiding being caught by authorities are dominant motivations. In *reactive* organizations, safety is beginning to be taken seriously, but only after an accident has already occurred. In *calculating* cultures, safety is maintained by various administrative systems and is primarily an issue imposed on employees. Some extent of labor force involvement in the safety work characterizes *proactive* cultures; in the most advanced type of safety culture (*generative*), safety is everyone's responsibility. The focus on safety is an important part of how business is conducted, and, even though there are few accidents or incidents, people do not relax (and rest on their laurels), but remain alert to dangers.

An advanced safety culture may also be described, according to Hudson (2003), by the four elements shown in Figure 7.1:

- *Information* means that all employees are informed and that information is shared within the organization. Importantly, bearers of bad news are not blamed. Instead, employees are encouraged to report problems.
- *Trust* is developed through treating employees in a fair way and not punishing those who report errors and mistakes.

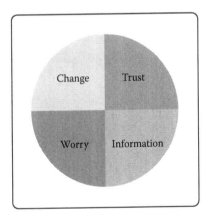

FIGURE 7.1 Elements that characterize a sound safety culture.

- *Change* describes such organizations. This means the organization is adaptable and learns from the times when something goes wrong and the times when something works well.
- *Concern or worries* imply a type of constant apprehensiveness and understanding of the fact that even though all precautions are made, something may go wrong.

7.7.3 STUDIES OF SAFETY CULTURE IN AVIATION

Several questionnaires have been developed to map safety culture in aviation; most are multidimensional and consider many of the aspects mentioned in the previous section. A study of ground personnel in a Swedish airline surveyed nine aspects, including communication, learning, reporting, and risk perception (Ek and Akselsson 2007). In the survey, managers were asked to reflect on what they thought the employees' responses would be. Managers were more positive about safety culture than operators. Meanwhile, significant consistencies were found for their evaluations of the various aspects of the company's safety culture. The lowest scores were given for the "justness" and "flexibility" dimensions; this applied to both management and employees. The former dimension assessed the extent to which making occasional mistakes was accepted, while the latter measured the extent to which employees were encouraged to provide suggestions for improvements. The highest scores were given for the dimensions "communication" and "risk perception" (Ek and Akselsson 2007).

7.8 WOMEN AND AVIATION

Figures from the United States reveal that about 90% of today's pilots are white males, which means that both females and other ethnic groups are underrepresented in the cockpit. There is reason to believe that this could change in the future job market, but the rate at which it would happen is uncertain. Naturally, it is difficult to predict how the industry will change over time and what the need for various professions will be in the years to come. Several Norwegian companies have expressed concern about a future shortage of pilots, and numbers from the United States reveal an expected increase of almost 27% in the demand for commercial pilots in the period to 2010 compared to the number employed in 2000 (U.S. Department of Labor, Bureau of Labor Statistics, quoted in Turney and Maxant 2004).

Thus, it is reasonable to believe that airlines will recruit pilots and personnel outside national borders and in other groups than those who have chosen the piloting profession traditionally. It is difficult to know exactly what motivates young people in their choice of profession. Probably, a number of factors are involved, such as subjective considerations of abilities and interests as well as external factors such as available opportunities, financial situation, and what they are familiar with through family and friends. Sometimes, even personal experiences may play a part in staking out a career path, as was the case with one of the first Norwegian female aviators, Gidsken Jacobsen (Gynnild 2008):

A day in June 1928, a big, three-engine seaplane roared trough Ofotfjorden and landed outside Narvik. Somewhere in the sea of spectators was Gidsken Jakobsen. The visit from above became a turning point in her life. "From the day I saw Nilsson's [flying] machine at the docks, there was nothing I'd rather do than fly," she said years later. "Imagine flying around in the air like Nilsson and his crew, from place to place, to get to know the country from the air and awaking the interest in thousands of people for what they loved more than anything else: To fly!"

Gidsken Jakobsen was raised in Narvik in the 1920s. She was not quite like other girls; she learned how to drive cars and motorbikes at an early age. At the age of 21, she left for Stockholm to take her pilot's license at the Aero-Materiell flight school. Subsequently, she learned to fly seaplanes and, with the help of her father, bought a seaplane that was given the name *Måsen* (*The Seagull*) (Gynnild 2008). There is no doubt that Gidsken Jakobsen lived an untraditional and exciting life, with great firmness of action and lust for life. Other examples of female pioneers can be found in Norway and in other countries, such as Dagny Berger and Elise Deroche, who got their pilot's licenses in 1927 and 1910, respectively. Harriet Quimby crossed the English Channel in 1912 and Amelia Earhart crossed the Atlantic in 1932 (Wilson 2004).

Despite the fact that women were at the center of early aviation and that there were many female pioneers, the piloting profession of today is distinctly male dominated. Although it is difficult to predict with accuracy the global rate of female participation in this profession, it has been estimated at between 3 and 4% in Western countries (Mitchell et al. 2005).

It is difficult to say why more women are not fascinated by the act of flying. Probably, there are individual reasons and reasons based on the nature of aviation. Perhaps the profession is seen as particularly masculine because history is full of heroic achievements performed by men—"the right stuff." Thus, young females would rather choose other education paths and career opportunities. Perhaps the thought of entering a trade where one risks being isolated and perhaps lives with negative and sexist comments from colleagues and others is less than appealing. Has the aviation industry been interested in allowing women into the cockpit, and with which attitudes are female pilots met? How can a female applicant to this profession expect to be received, and how do gender issues affect interactions in the cockpit, which, traditionally, is occupied by two males? Research into interactions between pilots has long been concerned with communication and cultural differences and their impact on aviation safety. Still, research into gender issues in aviation, both in terms of attitudes toward female pilots and consequences for interactions between pilots, is insufficient.

7.8.1 ATTITUDES TOWARD FEMALE PILOTS

To provide information about attitudes toward female pilots, a study was initiated in South Africa, the United States, Australia, and Norway; female and male pilots were asked about how they regarded female pilots (Kristovics et al. 2006). The subjects were asked to express their opinions on a number of statements about female pilots. In addition, the survey presented an open question to which participants could

TABLE 7.3

Examples of Questions from Survey on Attitudes toward Female Pilots

Category	Sample Statements
Decision/leadership	Female pilots often have difficulty making decisions in urgent situations.
	Female pilots' decision-making ability is as good in an emergency situation as it is in routine flights.
Assertiveness	Male pilots tend to "take charge" in flying situations more than female pilots do.
	Male flight students tend to be less fearful of learning stall procedures than female students are.
Hazardous behavior	Male pilots are more likely to run out of fuel than female pilots are.
	Male pilots tend to take greater risks than female pilots do.
Affirmative action	Professional female pilots are only in the positions they are in because airlines want to fulfill affirmative action quotas.
	Flight training standards have been relaxed so that it is easier for women to get their wings.

express in their own words what they thought about the issues raised by the survey and the survey itself. The survey contained four sections of questions (examples are presented in Table 7.3). The participants responded by expressing their agreement or disagreement on a five-point scale. Results were then summarized for each dimension so that a high score represented a positive attitude and a low score indicated a negative attitude. Some of the questions thus needed to be reversed to combine the scores into a unified index.

A total of 2,009 pilots (312 females and 1,697 males) with an average age of 36 participated in the survey. There were different proportions of participants from the four different countries, with 53% from Australia, 28% from South Africa, 9% from the United States, and 10% from Norway. Results revealed gender-related differences for all four dimensions; that is, male pilots regarded female pilots in a more negative light than female pilots did. Differences were greatest for the statements in the categories "decision/leadership" and "affirmative action."

There were also differences between countries, as presented in Figure 7.2. The following results are based on male pilots only because the numbers of female pilots were very low in some participating countries. For three of the four dimensions, the Norwegian pilots were more positive than pilots from other countries, whereas for the dimension labeled "hazardous behavior" the situation was reversed. Statements in the latter category were of the type: "Male pilots tend to take greater risks than female pilots," and Norwegian pilots, to a lesser extent, agreed with these statements. This probably reflects the equality-mindedness of Norwegians, in which female pilots would not be viewed as being more careful or apprehensive than male colleagues. In summary, the results of the survey revealed that male Norwegian pilots were more positive about female pilots than their U.S., Australian, and South African counterparts. Perhaps this expresses stronger ideals of equality in Norway compared to the other countries involved in the investigation, which is in line with Hofstede's (1980) findings for Scandinavian countries.

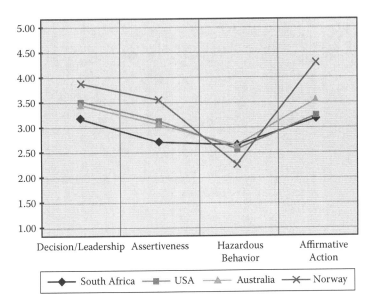

FIGURE 7.2   Differences in attitude in four countries based on male pilots' responses.

The group contained 158 pilot instructors. As a subgroup, instructors were found to be more positive toward female pilots compared to the group at large. Additionally, those who had had the opportunity to fly with a female co-pilot were more positive toward female pilots than those who had not.

Participants had the opportunity to provide supplemental comments to the survey, and there were several similarities in negative commentary from different countries (Mitchell et al. 2005). Some were about the aviation industry not being suited for women, exemplified in statements such as: "A female pilot = an empty kitchen" and "If women were meant to fly, the sky would be pink." The least amount of negative commentary was from the Norwegian group. Positive comments were often about how female pilots were just as good as male pilots, or better, and that it was unreasonable to judge people solely based on gender. Many Norwegian respondents (most of whom were military pilots) commented that because both males and females went through the same selection process, they were equally capable of performing the given tasks.

7.9   REORGANIZATION AND ADAPTING TO NEW WORKING CONDITIONS

Many industries are subject to restructuring and reorganization. Such changes are often attributed to a need or desire to improve the company's efficiency and profitability. Sometimes, economic conditions or strategic decisions lead to downsizing. However, one problem with downsizing is that the tasks to be performed do not disappear with the sacked staff; consequently, a greater workload rests on those who remain employed in the company. In addition, new technologies are often introduced, so employees must be willing to learn new skills and perform new tasks. Industries that were previously regarded as firmly established and able to offer attractive working conditions are no longer plentiful, leading to more shortsighted planning by employers and employees.

Restructuring is regarded as a common and correct strategy that, according to McKinley and Scherer (2000), provides managers with a feeling of cognitive order—a positive emotion that increases the probability of further restructuring. Although organizational changes are usually carried out with the best intentions, studies point to negative consequences in the form of stress and reduced work satisfaction among employees, as well as lower than expected financial gains for the company (for a review, consult Mentzer 2005). One interpretation of such outcomes involves insufficient and inappropriate management of human factors—in other words, underestimating the impact that expansive restructuring processes have on employees.

7.9.1 REACTIONS TO ORGANIZATIONAL CHANGES

For employees, consequences of corporate reorganization may be considerable. Work is a very important part of most people's lives, and significant changes in a company can be perceived as a critical life event. What individuals are most concerned about will vary, but may include questions such as

- How will this influence my day-to-day work routine and my career?
- Will my skills and experience be adequate in the future?
- Will I be in demand in the new organization?
- Will I be treated fairly in the reorganization process?

Restructuring may be seen in light of general stress theories—for example, Frankenhaeuser's bio-psychosocial model, which was mentioned in Chapter 6. In such models, restructuring would be regarded as a stressor to which individuals compare their resources. If demands surpass resources, stress arises and, in turn, negative health implications. What makes such a comparison or evaluation difficult is the scarcity of information about what will happen (and when). In the case of mergers and acquisitions, employees are usually not informed until relevant decisions have been made. Decision-making processes are typically secret on the grounds of protecting company interests. Internal rumors about what is about to happen often circulate, and this only makes it more difficult for employees to determine what will be required of them in the future and which tasks will be faced in the new organization. It is common not to know whether one has a job in the "new" company. That said, not everyone would react in a negative way to the message of an imminent restructuring or reorganization. Some people may consider such changes natural and appropriate, with opportunities for them to "climb the ladder."

In addition to individual reactions, what are the consequences of restructuring and downsizing to the organization? Initially, a more productive and dynamic organization is desired, but this outcome cannot be taken for granted. Potential savings must be weighed against possible negative consequences, such as increases in absences due to illness, higher turnover rates, and organizational unrest. Restructuring—in particular, downsizing—leads to reduced trust in management and lower work satisfaction among employees. How strong employee reactions will be depends on a number of circumstances. An important factor is the extent of the changes. The merger of two large organizations and the fusion of two smaller departments within an organization are different, for example. A decisive factor, though, is how the changes are communicated—notably, that the necessity of change is emphasized. Good communication contributes to reductions in anxiety and negative reactions, as well as improved work satisfaction in employees.

A study into reactions in Canadian hospital employees ($N = 321$) compared reactions to various forms of organizational changes including structural changes, workplace changes, and introduction of new technology (Bareil, Savoie, and Meunier 2007). Direct comparisons in how subjects reacted were possible because the subjects went through the same changes. A certain proportion (about a quarter) reacted with the same degree of discomfort to all the different types of changes, while the majority (about three quarters) experienced varying degrees of discomfort according to the type of change. Thus, it appears the type of change (situational factors), rather than personal characteristics (dispositional factors), is more important in terms of the level of discomfort associated with the changes.

In a Norwegian study into the restructuring and downsizing of a large oil company operating off the coast of Norway, employees' ($N = 467$) attitudes were investigated using a questionnaire (Svensen, Neset, and Eriksen 2007). The survey was conducted after the announcement (but before the implementation) of restructuring plans. About a third responded positively to the changes. The most important factors predicting a positive attitude were feelings of responsibility to the organization, involvement, participation, team leadership, and efficiency.

7.9.2 Downsizing

Downsizing has a number of negative implications for both those who lose their jobs and those who remain with the organization. Further, several studies show business earnings rarely improve as a consequence of downsizing (Mentzer 1996). Those who remain in the organization may suffer from so-called "survivor sickness"—a condition characterized by job security issues and negative or cynical attitudes toward the company. Even managers charged with executing reorganization and downsizing processes may experience stress, feelings of guilt, and, occasionally, aggression toward employees.

A study of employees in different sectors (including banking, insurance, and technology) investigated reactions in the wake of downsizing (Kets de Vries and Balazs 1997). The study was based on qualitative interviews of employees who lost their jobs ($N = 60$) and employees who remained with the company, so-called "survivors" ($N = 60$). Subjects in the former group were categorized according to their reactions: (a) the adapting (43%), (b) the depressed (30%), (c) those who see new opportunities (17%), and (d) the hostile (10%). In addition, combinations of reaction patterns were observed. Managers who had had to fire people were also subject to the study, and most of them said it was a difficult process. Great variations in the way managers reacted, ranging from distancing themselves from the problem to depression and feelings of guilt, were reported.

A survey conducted by Mari Rege at the University of Stavanger (in collaboration with Statistics Norway) tracked a large group of employees over 5 years (Rege, Telle, and Votruba in press). Employees in stable and secure companies were analyzed and compared to employees in companies exposed to downsizing. Downsizing was found to have a number of health-related consequences for the individual, particularly males. In males, results revealed higher mortality rates (14%), inability to work (24%), and risk of divorce (11%). The researchers attributed the gender-related differences to males taking job loss more seriously because male identities are more closely tied to their paid work. The study also revealed that the negative effects on someone who had lost his or her job depended on the sector in which he or she worked. The most severe consequences were observed in manufacturing industries.

7.9.3 Psychological Contracts

When seen as isolated or separate incidents, the reactions that managers and employees display in the previously described situations may seem strange, irrational, and even surprising. However, reactions become more predictable when viewed as reactions to intense stress. Both sides have been through a very difficult process that included losing one or more colleagues and, perhaps, friends. Moreover, they have been forced to change their perception of having worked in a financially solid company.

Downsizing may be regarded as a serious breach of the psychological contract that exists between employees and employers. Psychological contracts are individual expectations shaped by the organization about how exchanges should take place between the employee and the organization (see, for example, Rousseau 1995). Contracts are expectations about the future that are based on trust, acceptance, and reciprocity; they make it easier for people to plan and anticipate future events. Often these contracts contain the contributions expected of the employee as well as the compensation offered by the organization in return. For example, the employee promises to work hard, be loyal, and contribute to fulfilling the company's mission statement. The employer, on the other hand, promises ongoing employment, payment, and opportunities for personal development and career advancement. Such contracts are not necessarily synonymous with written contracts, and there may be certain disagreements about what the contract involves.

Psychological contracts are formed in many ways—for instance, through verbal expression, written documents, observation of how others are treated within the organization, and company policy or culture. They may be expressed in writing or by stories and myths about how things have been done before. Because contracts are formed by the person perceiving and integrating information, there is an obvious chance that misinterpretations can occur. Breach of contract arises when an employee feels that the organization has not fulfilled its duties. However, the organization or local management may have a different view.

Studies of people starting out in a new job show that contracts are often breached. One study revealed that 54% experienced a breach of contract during the first 2 years

(Robinson and Rousseau 1994). Thus, breaches of psychological contracts are not unusual in an organization. The severity of the breach will affect the severity of reactions and consequences. Breaches may occur knowingly and intentionally or because the business does not have the necessary resources to meet contractual requirements. Often, severe breaches of contract can be devastating and lead to negative reactions in employees, such as mistrust, anger, and wanting to quit the job. Generally, repeated offenses may degrade the relationship between the employee and the organization (Robinson and Rousseau 1994).

A number of factors influence the severity and nature of the consequences of breach of contract. One aspect is whether the breach occurred on purpose and whether similar breaches have occurred previously (a "string" of breaches). If an employer is unable to keep his or her promise to provide sponsored education because of a budget deficit, the employee may find it more acceptable than if the manager simply thinks such training is a bad investment. Events in the aftermath of a contractual breach may help to repair the relationship or, conversely, enforce its negative consequences.

7.10   LEADERSHIP

Sound management is important in many aspects of the working environment, particularly in relation to the development of a safety culture and in terms of processes of organizational change. A number of theories describe various leaders or leadership types. Leadership, naturally, does not exist in a vacuum but in a historical and cultural context; what works in one situation may not automatically transfer to a different organization or to a different point in time. In a time when many organizations are constantly changing and reinventing themselves, enormous demands are placed on managers to motivate and inspire employees. Many studies have shown a correlation between stress (burnout) and various forms of support or, perhaps, lack of support from managers (Lee and Ashforth 1996). Thus, it is crucial for an organization to be able to select and develop good leaders who earn and maintain employees' trust.

7.10.1   Three Leadership Types

Many theories and research traditions describe what makes a good leader. Some emphasize a leader's personality characteristics and others focus on what he or she actually does. A number of studies on leadership were conducted at Ohio State University in the postwar era. Two dimensions were identified as characteristic of effective leaders: that they were considerate and that they took initiative in generating structure (see, for example, Judge, Piccolo, and Ilies 2004). Other research groups have identified similar dimensions, but with different denotations, such as "relation-oriented leadership" and "task-oriented leadership."

Many new publications and books have been written recently on transactional leadership and transformational leadership (Burns 1978; Bass 2007). These leadership theories involve aspects concerning managers, subordinates, and the interactions between these two sides:

- In *transactional leadership,* the exchange of rewards for results and completion of tasks is emphasized. Alternatively, the employees are allowed to do their jobs as long as production targets are achieved; that is, error correction is considered sufficient. This may involve a passive, apprehensive style (the manager avoids action until something goes wrong) or an active style (the manager reacts to errors made by employees).
- In *transformational leadership,* emphasis is not only on transactions between employees and management (work for payment and other benefits), but also on the leader's ability to inspire, motivate, and devise original ideas. The person must be charismatic, set a good example, and be able to communicate his or her vision, thus elevating employees toward the organization's common goal. This leadership type is also described according to its impact on employees. Transformational leadership is associated with increased work satisfaction and motivates employees to perform better.

These two leadership styles are not mutually exclusive. Rather, they complement each other. A manager–employee relationship often starts as a transactional relation—that is, a process of clarifying the expectations that both sides have. However, transformational leadership is necessary if employees are to be motivated to put in additional effort (Bass 2007). Transformational leaders motivate employees to work not only for immediate rewards in self-interest, but also for the benefit of the group, the company, or, indeed, the country. Work becomes an activity greater than something done just to get paid, and this recognition contributes to an increase in employees' self-esteem and devotion to their tasks.

Evidence tells us that this categorization of leadership is not tied to a particular type of organization or culture, and the dimensions have been established in a number of organization types, such as the armed forces and the private and public sectors of many countries (Bass 2007). However, the ways in which the different forms of leadership are revealed may vary between cultures. For example, the way a leader rewards or shows appreciation for employees will vary between countries such as Norway and Japan.

The two forms of leadership are often contrasted to laissez-faire leadership, or a lack of leadership. Managers who exhibit such leadership do not recognize their responsibility as leaders and do not provide assistance or communicate their opinions on important issues regarding the organization. This form of leadership is the least effective and also represents the type of management under which employees are the least content with the status quo.

The different forms of leadership are often quantified using the measurement instrument called MLQ (multifactor leadership questionnaire) (Bass 1985). The instrument has subsequently been revised and consists, in short, of various statements on leadership to which employees respond on a scale from zero (the behavior is never observed) to four (the behavior is often, if not always, observed). Three scales map *transformational leadership* and, accordingly, three measure *transactional leadership.* However, only one scale measures *laissez-faire leadership.* A number of studies have used this instrument or variants of it in which the different scales are correlated with work performance measures (subjective and objective).

Summaries show that positive correlations are strongest for transformational scales and some of the transactional scales, whereas correlations between laissez-faire leadership and work results are negative (see, for example, Bass 2007). These findings are generally based on North American studies; however, a clear connection between transformational leadership and variables such as work satisfaction and efficiency has been supported in Norwegian surveys as well (Hetland and Sandal 2003).

Gender issues and leadership have been subject to plenty of discussion. For example, is it true that females have different leadership styles from those of males? In most industries, women are underrepresented in leadership positions, particularly at the highest levels. Do female leadership styles constitute a barrier to clinching those top jobs? A meta-analysis of 45 studies that investigated gender differences found generally minor differences in male and female leadership (Eagly, Johannesen-Schmidt, and van Engen 2003). Women had somewhat higher scores for transformational leadership, and men scored higher on the scales measuring laissez-faire leadership and the two forms of transactional leadership—that is, intervening only when something goes wrong. These results are positive for the case of female leadership. In other words, no evidence supports claims that female leaders use less effective leadership styles—quite the contrary.

With regard to the correlation between personality and leadership styles, several studies have been executed on the topic. A meta-analysis of over 20 studies conducted by Bono and Judge (2004), revealed low correlations between personality traits (the "big five" model) and the various scales presented in MLQ. The greatest and most stable correlations were found for the scales for transformational leadership and the personality traits extroversion (mean $r = .24$) and neuroticism (mean $r = -.17$).

7.10.2 LEADERSHIP AND SAFETY

Few studies on leadership in aviation have been conducted. However, a number of studies in other sectors may shed light on the relationship between leadership and safety. Several studies have investigated the connection of management, safety climate, and work-related accidents. A model for the relationship among these elements was developed by Barling, Loughlin, and Kelloway (2002). In this model, transformational leadership affects the security climate, which, in turn, affects the occurrence of accidents. The model also contained the important notion of safety awareness. Enforcement of this notion implies that employees are aware of the dangers associated with the various operations and know what needs to be done when accidents and dangerous situations arise. The model was tested on a group of young employees, mainly in the service sector. Results indicate that the most important effect of transformational leadership was its influence on safety awareness in individual employees; in turn, this affected the safety climate and, ultimately, the number of accidents (Barling, Loughlin, and Kelloway 2002).

Others have researched corresponding models for different sectors—for example, factory workers in Israel, where safety climate was found to be a mediating variable between leadership styles and the number of accidents (Zohar 2002). In this study, both transformational and transactional leadership predicted accidents; however, the effects were mediated through a certain aspect of the safety climate that was labeled "preventive action." This dimension contained questions that mapped the extent to which immediate superiors discussed safety issues with employees and whether they accepted safety advice from employees. The study concluded that leadership dimensions associated with concern for the welfare of employees and personal relations promoted improved supervision and thus improved the safety climate and reduced the number of accidents (Zohar 2002).

7.11 SUMMARY

For better or worse, people are influenced by the cultural context of groups or communities to which they belong. In this chapter, we have looked at how national, professional, and organizational cultures influence people working in aviation. Culture affects not only the behavior of a person, but also the way he or she communicates and perceives other people and the world (worldview).

Aviation is an inherently international industry that, similarly to many other industries, faces challenges in the form of tough competition, market instability, and increased focus on security and terrorism. Restructuring and downsizing are of great importance to the welfare of employees and, in turn, their performance and duties in relation to the organization. Sound leadership is of decisive importance to development of a safety culture and to reducing unintended consequences of restructuring and reorganization.

RECOMMENDED READING

Helmreich, R. L., and Merritt, A. C. 1998. *Culture at work in aviation and medicine.* Aldershot, England: Ashgate.
Maslach, C., and Leiter, M. 1997. *The truth about burnout.* San Francisco, CA: Jossey–Bass.

REFERENCES

Bang, H. 1995. *Organisasjonskultur* (3.utgave) [*Organizational culture*]. Oslo: Tano AS.
Bareil, C., Savoie, A., and Meunier, S. 2007. Patterns of discomfort with organizational change. *Journal of Change Management* 7:13–24.
Barling, J., Loughlin, C., and Kelloway, E. K. 2002. Development and test of a model linking safety-specific transformational leadership and occupational safety. *Journal of Applied Psychology* 87:488–496.
Bass, B. M. 1985. *Leadership and performance beyond expectations.* New York: Free Press.
———. 2007. Does the transactional–transformational leadership paradigm transcend organizational and national boundaries? *American Psychologist* 52:130–139.
Bono, J. E. and Judge, T. A. 2004. Personality and transformational and transactional leadership: A meta-analysis. *Journal of Applied Psychology* 89:901–910.
Burns, J. M. 1978. *Leadership.* New York: Harper & Row.
Eagly, A. H., Johannsesen-Schmidt, M. C., and van Engen, M. L. 2003. Transformational, transactional, and lassez-faire leadership styles: A meta-analysis comparing women and men. *Psychological Bulletin* 129:569–591.
Ek, Å., and Akselsson, R. 2007. Aviation on the ground: Safety culture in a ground handling company. *International Journal of Aviation Psychology* 17:59–76.

Gynnild, O. 2008. *Seilas i storm. Et portrett av flypioneren Gidsken Jakobsen* [*Sailing in storm: A portrait of the aviation pioneer Gidsken Jacobsen*]. Stamsund: Orkana Forlag og Norsk luftfartsmuseum.

Helmreich, R. L., and Merritt, A. C. 1998. *Culture at work in aviation and medicine.* Aldershot, England: Ashgate.

Hetland, H., and Sandal, G. M. 2003. Transformational leadership in Norway: Outcomes and personality correlates. *European Journal of Work and Organizational Psychology* 12:147–170.

Hofstede, G. 1980. *Culture's consequences: International differences in work-related values.* Beverly Hills, CA: Sage.

———. 2001. *Culture's consequences: Comparing values, behaviors, institutions and organizations across nations,* 2nd ed. Thousand Oaks, CA: Sage.

———. 2002. Commentary on "An international study of the psychometric properties of the Hofstede values survey module 1994: A comparison of individual and country/province level results." *Applied Psychology: An International Review* 51:170–178.

Hudson, P. 2003. Applying the lessons of high-risk industries to health care. *Quality and Safety in Health Care* 12:7–12.

Judge, T. A., Piccolo, R. F., and Ilies, R. 2004. The forgotten one? The validity of consideration and initiating structure in leadership research. *Journal of Applied Psychology* 89:36–51.

Kets de Vries, M. F. R., and Balazs, K. 1997. The downside of downsizing. *Human Relations* 50:11–50.

Kristovics, A., Mitchell, J., Vermeulen, L., Wilson, J., and Martinussen, M. 2006. Gender issues on the flight-deck: An exploratory analysis. *International Journal of Applied Aviation Studies* 6:99–119.

Lee, R. T., and Ashforth, B. E. 1996. A meta-analytic examination of the correlates of the three dimensions of job burnout. *Journal of Applied Psychology* 81:123–133.

Li, W. C., Harris, D., and Chen, A. 2007. Eastern minds in Western cockpits: Meta-analysis of human factors in mishaps from three nations. *Aviation, Space, and Environmental Medicine* 78:420–425.

McKinley, W., and Scherer, A. G. 2000. Some unanticipated consequences of organizational restructuring. *Academy of Management Review* 25:735–752.

Mearns, K. J., and Flin, R. 1999. Assessing the state of organizational safety—culture or climate? *Current Psychology* 18:5–17.

Mentzer, M. S. 1996. Corporate downsizing and profitability in Canada. *Canadian Journal of Administrative Sciences* 13:237–250.

———. 2005. Toward a psychological and cultural model of downsizing. *Journal of Organizational Behavior* 26:993–997.

Merritt, A. 2000. Culture in the cockpit. Do Hofstede's dimensions replicate? *Journal of Cross-Cultural Psychology* 31:283–301.

Mitchell, J., Kristovics, A., Vermeulen, L., Wilson, J., and Martinussen, M. 2005. How pink is the sky? A cross-national study of the gendered occupation of pilot. *Employment Relations Record* 5:43–60.

Mjøs, K. 2002. Cultural changes (1986–96) in a Norwegian airline company. *Scandinavian Journal of Psychology* 43:9–18.

———. 2004. Basic cultural elements affecting the team function on the flight deck. *International Journal of Aviation Psychology* 14:151–169.

Pidgeon, N., and O'Leary, M. 2000. Man-made disasters: Why technology and organizations (sometimes) fail. *Safety Science* 34:15–30.

Rege, M., Telle, K., and Votruba, M. In press. The effect of plant downsizing on disability pension utilization. *Journal of the European Economic Association.*

Robinson, S. L., and Rousseau, D. M. 1994. Violating the psychological contract: Not the exception but the norm. *Journal of Organizational Behavior* 15:245–259.

Rousseau, D. M. 1995. *Psychological contracts in organizations.* London: Sage Publications.

Schein, E. H. 1990. Organizational culture. *American Psychologist* 45:109–119.

———. 1996. Culture: The missing concept in organization studies. *Administrative Science Quarterly* 41:229–240.

Sherman, P. J., Helmreich, R. L., and Merritt, A. C. 1997. National culture and flight deck automation: Results of a multination survey. *International Journal of Aviation Psychology* 7:311–329.

Shouksmith, G., and Taylor, J. E. 1997. The interaction of culture with general job stressors in air traffic controllers. *International Journal of Aviation Psychology* 7:343–352.

Soeters, J. L., and Boer, P. C. 2000. Culture and flight safety in military aviation. *International Journal of Aviation Psychology* 10:111–113.

Spector, P. E., and Cooper, C. L. 2002. The pitfalls of poor psychometric properties: A rejoinder to Hofstede's reply to us. *Applied Psychology: An International Review* 51:174–178.

Svensen, E., Neset, G., and Eriksen, H. R. 2007. Factors associated with a positive attitude towards change among employees during early phase of a downsizing process. *Scandinavian Journal of Psychology* 48:153–159.

Turney, M. A., and Maxant, R. F. 2004. Tapping diverse talent: A must for the new century. In *Tapping diverse talent in aviation,* ed. M. A. Turney, 3–10. Aldershot, England: Ashgate.

Westrum, R., and Adamski, A. J. 1999. Organizational factors associated with safety and mission success in aviation environments. In *Handbook of aviation human factors,* ed. D. J. Garland, J. A. Wise, and V. D. Hopkin, 67–104. Mahwah, NJ: Lawrence Erlbaum Associates.

Wiegmann, D. A., and Shappell, S. A. 2003. *A human error approach to aviation accident analysis: The human factors analysis and classification system.* Burlington, VT: Ashgate.

Wiegmann, D. A., Zang, H., Von Thaden, T. L., Sharma, G., and Gibbons, A. M. 2004. Safety culture: An integrative review. *International Journal of Aviation Psychology* 14:117–134.

Wilson, J. 2004. Gender-based issues in aviation, attitudes towards female pilots: A cross-cultural analysis. Unpublished doctoral dissertation. Pretoria, South Africa: Faculty of Economic and Management Sciences, University of Pretoria.

Zohar, D. 2002. The effects of leadership dimensions, safety climate, and assigned priorities on minor injuries in work groups. *Journal of Organizational Behavior* 23:75–92.

8 Aviation Safety

8.1 INTRODUCTION

An American cowboy rodeo saying states, "There's never been a horse that can't be rode; there's never been a rider that can't be throw'd." That adage applies equally well to aviation. There has never been a pilot so skilled that he or she cannot have an accident. Clearly, some pilots are exceptionally skilled and cautious. However, given the right set of circumstances, even they can make an error of judgment or find that the demands of the situation exceed their capacity or the capabilities of their aircraft. What sets these particular pilots apart is that these combinations of events and circumstances occur very rarely; their attributes, including attitudes, personality, psychomotor coordination, aeronautical knowledge, skills, experiences, and a host of other individual characteristics, make them less likely to experience hazardous situations and more likely to survive the situations if they occur.

In contrast, for pilots at the low end of the skill continuum, every flight is a risky undertaking. In this chapter, we will explore some of the research that has attempted to explain, from the perspective of human psychology, how these groups of pilots differ, why accidents occur, and what might be done to reduce their likelihood.

8.2 ACCIDENT INCIDENCE

To begin, let us examine the incidence of aviation accidents so that we may understand the extent of the problem. Table 8.1 shows the numbers of accidents and corresponding accident rates (number of accidents per 100,000 flight hours) for a 2-year period in the United States. From this table, the differences in accident rates of the large air carriers (very low rates), the smaller carriers, and general aviation are evident. Over that span of operation, the accident rate increases about 30-fold. To put these statistics in a slightly different light, on a per-mile basis, flying in an air carrier is about 50 times safer than driving. However, flying in general aviation is about seven times riskier than driving.

These accident rates are typical of the rates found in Western Europe, New Zealand, and Australia. For example, data from the Australian Transport Safety Bureau (ATSB 2007) show fixed-wing, single-engine general aviation accident rates (accidents/100,000 hours) of 10.26 and 7.42 for 2004 and 2005, respectively. Note that these rates are somewhat inflated relative to the United States because they do not include multiengine operations normally used in corporate aviation, which is traditionally one of the safest aviation settings.

This brings up an important point that must be made regarding safety statistics. It is very important to note the basis on which the statistics are calculated. For example,

TABLE 8.1
Incidence of Accidents in the United States

	2004		2005	
	Number	**Rate**[a]	**Number**	**Rate**[a]
Large air carriers	30	0.16	39	0.20
Commuter	4	1.32	6	2.00
Air taxi	66	2.04	66	2.02
General aviation	1617	6.49	1669	6.83

Source: Federal Aviation Administration. 2007.

[a] Rate is given as accidents per 100,000 flight hours.

in Table 8.1, the rates are given in terms of numbers of accidents per 100,000 flight hours. This is a commonly used denominator, but by no means the only one that is reported. Our earlier comparison of accident risk in driving and aviation used accidents per mile traveled. Some statistics are in terms of numbers of departures (typically, accidents per one million departures). It is important for the reader to make note of these denominators so that comparisons are always made between statistics using the same denominator. In addition, as in our comparison between the statistics from the United States and Australia, it is important to know exactly what has been included in the calculations. In this case, exclusion of the very safe, multiengine corporate operations could lead to the conclusion that general aviation is safer in the United States than in Australia—a conclusion that is not warranted by the data provided.

8.3   CAUSES OF ACCIDENTS

For every complex question, there is a simple answer—and it's wrong.

Attributed to H. L. Mencken

Before we begin to talk about the causes of accidents, we need to make clear what we mean by a "cause." Step away from the flight line for a moment and into the chemistry laboratory. If we were to put a few drops of a solution containing silver nitrate ($AgNO_3$) into another solution that contains sodium chloride (NaCl, common table salt), we would observe the formation of some white particles (silver chloride, AgCl) that would sink to the bottom of our test tube. This simple test for the presence of chlorine in water by the addition of aqueous silver nitrate is, in fact, one of the most famous reactions in chemistry, and it is among the first learned by all budding chemists. The point to be made here is that this reaction and the formation of the precipitate will happen every single time that we mix solutions of silver nitrate and sodium chloride. The precipitate will not form unless we add the silver nitrate. The addition of the silver nitrate to the sodium chloride solution is a necessary and sufficient condition for the formation of the precipitate. We may truly say that one causes the other.

Now step back outside the laboratory and consider what happens in the real world. For example, let us imagine an individual driving to work one morning when traffic is very heavy; he is following closely behind the vehicle ahead. Occasionally, that vehicle will brake sharply, so he needs to react quickly and apply the brakes to keep from hitting it. This happens dozens, perhaps hundreds, of times during the trip and he is always successful in avoiding an accident. During the same trip, he listens to music on the radio and occasionally changes the station by glancing at the radio and pressing the buttons to make a selection. He may do this several times during the course of the trip, also without incident. There may even be occasions when, as he is changing stations on the radio, the vehicle ahead brakes, and he glances up just in time to notice the brake lights and slow down. Fortunately, he is a careful driver and usually maintains an adequate spacing between his vehicle and the vehicle he is following, so he is always able to react in time, even if he is temporarily distracted by the radio. He may do this every day for years, without incident.

However, on one particular morning he is delayed leaving the house, so he does not get his usual cup of coffee and is feeling a little sleepy. He is also feeling a bit rushed because he needs to be at the office at the usual time, but he has gotten a late start. Perhaps this has led him to follow the vehicle ahead a little more closely than usual and, now, as he reaches over to change the radio, the driver ahead brakes more sharply than usual; he does not notice the vehicle's brake lights quite soon enough or react quickly enough to slow his vehicle. An accident occurs. But what was the cause of the accident?

From the official standpoint (the one that will go on the police report), the individual in the vehicle behind the braking driver was the cause, and this is yet another example of human error. However, that is not a very satisfying explanation. It is not satisfying because it describes actions taken on almost every trip for many years as an error. Surely, there have been many days on which the individual left the house late and hurried to make up time. Surely there have been days when he felt a little sleepy when driving to work. Likewise, he has handled heavy traffic and changing radio stations innumerable times previously. All of these actions and conditions have existed previously and we have not called them errors and the causes of an accident because, until this particular day, no accident has occurred. None of these conditions and events is necessary and sufficient for an accident to occur. However, each of them, in its own small way, increased the likelihood of an accident.

Therefore, we suggest that the best way to understand the causes of accidents is to view them as events and conditions that increase the likelihood of an adverse event (an accident) occurring. None of the usual list of causes—following too closely, inattention, sleepy driver, distraction—will cause an accident to occur each and every time it is present. However each will independently increase the likelihood of an accident. Moreover, their joint presence may increase the likelihood far more than the simple sum of their independent effects. For example, following too closely in traffic and driving while drowsy both increase the risk of an accident—let us say by 10% each. However, following too closely in traffic *while* drowsy might increase the risk of an accident by 40%, not the 20% obtained by simply summing their independent contributions. Thus, the combination of these two conditions is far more dangerous than either is by itself.

Causes are best understood as facilitators of accidents rather than as determinants of accidents. They increase the probability that an accident will occur, but they do not demand that it occur. This argument implies that accidents generally have multiple facilitating components (causes).

Most authors, at least in recent years, acknowledge in the introduction to their research that there is no single cause for accidents and then proceed to ignore that statement in the conduct and interpretation of their research. Arguably, the present authors could be included in that indictment. However, to atone for those past literary indiscretions, let us now reiterate that point: *There are no single causes for accidents.*

Usually the "cause" is simply the last thing that happened before the crash. As this was being written, an Airbus was being extracted from the Hudson River after both engines failed at 3,200 feet during takeoff from LaGuardia Airport. The newspapers reported that the *cause* of the crash was the engines' ingestion of a flock of geese. However, they also reported that the captain of the flight was an experienced glider pilot, with an exceptional interest in safety. Clearly, multiple *causes* were at work here—the flock of geese may have *caused* the engines to quit, but the experience and skill of the captain may have been the *cause* of the relatively benign water landing that resulted in no fatalities.

In exploring cause-and-effect relationships, we may move away from the final cause to whatever extent results in a comprehensive understanding of the event. For example, we might ask what caused the geese to be in the flight path of the aircraft. Did placing a major airport along a river in the flyway for migratory waterfowl play some part? We might also ask what part the pilot's gliding experiences played in the outcome. Did they "cause" a catastrophic event to become an exciting, but injury-free event? When we take a more situated view, we recognize that there are no "isolated" events. Everything happens in a context.

Each accident occurs because of a complex web of interacting circumstances, including environmental conditions, pilot attributes, aircraft capabilities, and support system (e.g., air traffic control, weather briefer) weaknesses. A complete explanation of how those elements interact to produce an accident is far beyond our current science. Science does not, at this time, allow us to predict with anything approaching certainty that, under a well-specified set of circumstances, an accident will occur; this is definitely not the chemistry laboratory.

To begin, we do not know the set of circumstances that should be specified or the values to assign to the various elements so that they combine properly. Despite this abundant ignorance, we are able to make some statements regarding probabilities. That is, we are able to say with some confidence that accidents are more likely to occur under some circumstances than under other circumstances. The identification of these circumstances and the establishment of the degree of confidence with which we may assert our beliefs make up the topic to be considered next.

Many efforts have been made to identify the causes for aircraft accidents over the years. Although they suffer from the implicit assumption of single causes, which we have dismissed as naive, these efforts nevertheless can make a contribution to our understanding of accident causality by identifying some of the circumstances and attributes associated with accidents.

To look at accident causality in a slightly different way, consider the following parable.

Imagine you are standing in a calm pool of water. You reach into your pocket and grasp a handful of pebbles and toss them into the pool. Each of the pebbles disturbs the surface of the pool, creating an exceptionally complex set of interacting waves. Because some of the pebbles are larger than others, their waves are higher than those of the smaller pebbles. Where the waves intersect, they combine algebraically, depending on the phase and magnitude of each individual wave. At some points, they combine to produce a wave that reaches high above the surface. At other points, they cancel each other out. Occasionally, several of the individual waves will intersect at just the right moment to produce a freak wave of exceptional height that will cause the water to lap over the tops of your boots—an adverse event. For all intents and purposes, the occurrence of this freak wave is random and largely unpredictable. Its production depends on the number of pebbles plucked from your pocket, the size of the pebbles, the height of their toss into the pool, and the amount of dispersion of the pebbles as they fall down into the water. It will also depend upon who else is standing in or around the pool tossing pebbles into the water.

Observation of many tosses of pebbles into the pool and the use of statistics will allow us to predict that a freak wave will occur every so often—let us say once in every 100 tosses. However, we cannot reliably say whether any particular toss will produce the freak wave or where on the surface of the pool this wave will occur. Even so, we are not powerless to prevent its occurrence. We might, for instance, reduce the number of pebbles we throw. We might reduce the size of the pebbles. We might change the vigor with which we throw them into the air. We might throw them less vertically and more laterally so as to increase their dispersion. We might also make rules about how often the people on the shore can throw pebbles into the pond.

Freak waves might still occur, but now we might only see them once in every 1,000 tosses. Further reductions might be achieved by, for example, coating the pebbles with some substance that reduces their friction as they pass through the surface of the water, hence producing individual waves of still lower magnitude. However, even with all these procedural and technological advances, as long as we toss pebbles into the pool, there is some nonzero chance that a freak wave will occur. Completely eliminating freak waves (and the attendant adverse event—wet feet) requires that we and our companions abandon our practice of tossing pebbles into the pool or that we take a radically different approach. For example, we might wait for winter, when the pool will freeze.

8.4   CLASSIFICATION OF AIRCRAFT ACCIDENTS

According to the Aircraft Owners and Pilots Association (AOPA 2006), causes of accidents may be broken down into three categories:

- *Pilot-related* accidents arise from the improper action or inaction of the pilot.

TABLE 8.2

Causes of General Aviation Accidents in 2005

Major Cause	All Accidents		Fatal Accidents	
Pilot	1076	74.9%	242	82.9%
Mechanical/maintenance	232	16.2%	22	7.5%
Other/unknown	128	8.9%	28	9.6%
Total	1436		292	

- *Mechanical/maintenance* accidents arise from failure of a mechanical component or errors in maintenance.
- *Other/unknown* accidents include causes such as pilot incapacitation, as well as accidents for which a cause could not be determined.

Table 8.2 shows the distribution of accidents among those three categories of cause for general aviation accidents in 2005. Clearly, the predominant major cause for both levels of severity was the pilot. Considering only those accidents in which the pilot was the major cause, the AOPA further divided the accidents among the categories shown in Table 8.3. Interpretation of the data presented in Table 8.3 is made difficult by the mixture of stage-of-flight categories (i.e., preflight/taxi, takeoff/climb, and landing) with two categories (fuel management, weather) that are conceptually unrelated to the other stage-of-flight categories. This admixture of taxonomic elements muddies the interpretation of an analysis of only dubious initial value. At most, one

TABLE 8.3

Accident Categories for Pilot-Related Accidents

Category	Total		Fatal	
Preflight/taxi	38	3.5%	1	0.4%
Takeoff/climb	165	10.5%	33	13.6%
Fuel management	113	10.5%	20	8.3%
Weather	49	4.6%	33	13.6%
Other cruise	21	2%	14	5.8%
Descent/approach	49	4.6%	25	10.3%
Go-around	43	4.0%	15	6.2%
Maneuvering	122	11.3%	80	33.1%
Landing	446	41.4%	8	3.3%
Other	30	2.8%	13	5.4%

Source: Aircraft Owners and Pilots Association. 2006. *The Nall report,* p. 8. Frederick, MD: Author.

might inspect these data and conclude that maneuvering flight is a dangerous phase. However, these data say nothing about why maneuvering flight is dangerous or demonstrate that it is relatively more dangerous than other stages of flight because there is no control for exposure—the amount of time spent in that flight stage.

We belabor this point to reinforce the notion that categories are not causes. Categories do not explain why accidents occur. They simply point to times, conditions, and circumstances under which accidents are more likely. To illustrate this point one last time, an accident does not occur simply because the pilot was in maneuvering flight. It occurred while the pilot was in maneuvering flight *and* decided to buzz his friend's house *and* was distracted *and* flew too slowly *and* stalled the aircraft *and* encountered a downdraft *and* was flying an underpowered aircraft *and* had too much load on board *and*…—the list could go on for a very long time. The reader should recall our earlier discussion about how each of these "causes" increases the likelihood of an accident while not guaranteeing its occurrence.

In a seminal and frequently cited study, Jensen and Benel (1977) noted that all aircrew errors could be classified into one of three major categories based on behavioral activities: procedural, perceptual-motor, and decisional tasks. This conclusion was based on an extensive review of all U.S. general aviation accidents occurring from 1970 to 1974 using data from the National Transportation Safety Board (NTSB). Of the fatal accidents involving pilot error during that period, Jensen and Benel found that 264 were attributable to procedural errors, 2,496 had perceptual-motor errors, and 2,940 were characterized as having decisional errors. Examples of procedural tasks include management of vehicle subsystems and configuration; related errors would include retracting the landing gear instead of flaps or overlooking checklist items. Perceptual motor tasks include manipulating flight controls and throttles, and errors would include overshooting a glide-slope indication or stalling the aircraft. Decisional tasks include flight planning and in-flight hazard evaluation; errors would include failing to delegate tasks in an emergency situation or continuing flight into adverse weather.

Diehl (1991) analyzed U.S. Air Force and U.S. civil air carrier accident data for accidents that occurred during 1987, 1988, and 1989. His analysis of the air carrier data indicated that 24 of the 28 major accidents (those resulting in destroyed aircraft and/or fatalities) involved aircrew error. Of these accidents, 16 procedural, 21 perceptual-motor, and 48 decisional errors were cited (the errors sum to more than 24 because some accidents involved multiple errors). During the same time period, 169 major mishaps were reported for the U.S. Air Force; these involved destruction of the aircraft, over one million dollars in damages, or fatalities. Of the 169 major mishaps, 113 involved some type of aircrew error. These included 32 procedural, 110 perceptual-motor, and 157 decisional errors. These types of errors were labeled "slips," "bungles," and "mistakes," respectively (Diehl 1989).

The comparison of the incidence of these three types of errors among the three aviation sectors is depicted in Table 8.4. It is interesting to note that even though these three sectors differ vastly with regard to many aspects (e.g., training, composition of aircrew, type of aircraft, and type of mission), the relative incidence of the errors is remarkably similar for all three groups.

TABLE 8.4
Types of Aircrew Errors in Major Accidents

Kind of Operation	Category of Error		
	Procedural "Slips"	Perceptual-Motor "Bungles"	Decisional "Mistakes"
General aviation	5%	44%	52%
Airlines	19%	25%	56%
Military	11%	37%	53%

Source: Adapted from Diehl, A. E. 1991. Paper presented at the 22nd International Seminar of the International Society of Air Safety Investigators. Canberra: November 1991.

The analyses discussed so far were largely conducted on an ad hoc basis. That is, they were not conducted on the basis of a specific, well-defined theory of why accidents occur. However, the work of Perrow (1984) on the nature of accidents in closely coupled systems and the work by Reason (1990, 1997) have led to one theoretical conceptualization of why accidents occur. This theory, most widely articulated by Reason, is commonly referred to as the "swiss-cheese model" and suggests that governments, organizations, and people create barriers to the occurrence of accidents. Examples of barriers include regulations that prescribe certain rest periods between flights, checklists that must be followed during the planning and conduct of a flight, procedures for the execution of an approach to landing, and standard operating procedures for resolving normal and abnormal situations.

Each of these barriers is created to prevent or require behavior that results in safe operations. As long as the barrier is in place and intact, no errors associated with that barrier can occur. The barrier acts as a shield to prevent accidents from occurring or to prevent the effects of failures of other barriers from resulting in an accident. One might envision these barriers as stacked one against the other. However, no barrier is perfect. Each might be described as having "holes"—areas in which the defenses associated with a particular barrier are weak or missing—hence, the swiss-cheese description, as depicted in Figure 8.1.

The swiss-cheese model of accident causation proposed by Reason was operationalized by Wiegmann and Shappell (1997, 2003), who developed a system for categorizing accidents according to the sequential theory proposed by Reason. Wiegman and Shappell termed this approach, which is a taxonomy that describes the human factors that contribute to an accident, the human factors analysis and classification system (HFACS). The system has four levels arranged hierarchically. At the highest level are the organizational influences. Next are aspects of unsafe supervision, followed by the preconditions for unsafe acts. Finally, at the lowest level, are the unsafe acts of operators. These levels correspond with the barriers to accidents shown in Figure 8.1. These major components may be further broken down into the following elements (Shappell and Wiegmann 2000):

- organizational influences
  - resource management

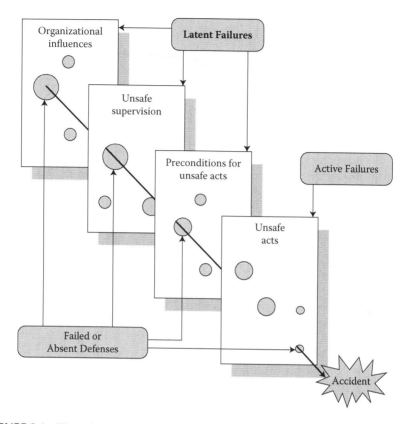

FIGURE 8.1   The swiss-cheese model of accident causation.

- organizational climate
- organizational process

- unsafe supervision
  - inadequate supervision
  - planned inappropriate operations
  - failure to correct a problem
  - supervisory violations

- preconditions for unsafe acts
  - substandard conditions of operators
    - adverse mental states
    - adverse physiological states
    - physical and mental limitations
  - substandard practices of operators
    - crew resource management
    - personal readiness

- unsafe acts of operators
  - errors
    - skill based
    - perceptual
    - decisional
  - violations
    - routine—habitual departures from rules sanctioned by management
    - exceptional—departures from rules not sanctioned by management

HFACS has been used in a variety of settings, including analyses of accidents in the U.S. military services and analyses of accidents among air carriers and general aviation in the United States (e.g., Shappell and Wiegmann 2002, 2003; Shappell et al. 2006). It has also been used in settings outside the United States (e.g., Gaur 2005; Li and Harris 2005; Markou et al. 2006).

In a study conducted by the Australian Transport Safety Bureau, Inglis, Sutton, and McRandle (2007) conducted a comparison of U.S. and Australian accident causes using HFACS. The results of their analyses are shown in Table 8.5. The authors concluded:

> The proportion of accidents that involved an unsafe act was similar: of the 2,025 accidents in Australia, 1,404 (69%) were identified as involving an unsafe act while 13,700 accidents (72%) in the US involved an unsafe act. Moreover, the pattern of results between United States and Australian accidents was remarkably similar. The rank order of unsafe act categories was the same in the accident sets for both countries. Skill-based errors were by far the most common type of aircrew error followed by decision errors, violations and perceptual errors, in that order. (p. 39)

Although HFACS has achieved widespread use, it is not without its limitations, critics, and alternatives. For example, it is important to note that HFACS is a secondary analysis process. That is, HFACS does not deal with primary data. Rather, it

TABLE 8.5
Accidents Associated with Each HFACS Unsafe Act

Unsafe Act	Australia		United States	
	Frequency	%	Frequency	%
Skill-based error	1,180	84	10,589	77.3
Decision error	464	33	3,996	29.2
Perceptual error	85	6.1	899	6.6
Violation	108	7.7	1,767	12.9
Sample size	1,404		13,700	

Source: Inglis, M., Sutton, J., and McRandle, B. 2007. Human factors analysis of Australian aviation accidents and comparison with the United States (aviation research and analysis report—B2004/0321). Canberra: Australia Transport Safety Bureau.

deals with data generated by accident investigators and is in that sense removed from the reality of the accident. In contrast, consider a botanist classifying a new leaf. The botanist examines the leaf itself—the shape, arrangement of veins, coloration, and perhaps even the chemical composition of the leaf. The botanist does not examine a narrative description of someone else's impressions of a leaf. Yet, this is precisely the case with HFACS. The analysts using HFACS use the narrative report of an accident prepared by an accident investigator as their data. Although studies (Shappell et al. 2006; Wiegmann and Shappell 2001) have shown that these analysts exhibit good reliability in their judgments, the results are still subject to the validity, or lack thereof, of the original accident investigator. Further, the analysts are largely confined to examining only those data that the original investigator considered relevant and/or those data required by the regulatory body responsible for the investigation.

At a more basic level, the use of "disembodied data," to use Dekker's (2001) phrase, to explain a complex event after the fact is arguably a futile endeavor. It is only with hindsight that we are able to label decisions as poor judgment or actions as mistakes. Taken within the context of the situation, these decisions and actions may well have seemed entirely appropriate and correct to the individuals at the time, given the knowledge they possessed. Further, the utility of assigning accidents to categories is not entirely clear. Knowing that an accident belongs to a specified category does not imply that we know why the accident occurred. To know that an accident occurred because of a "skill-based pilot error" does not tell us why what we have labeled as an error after the fact occurred. Labeling is not the same as understanding. Labeling will never lead to prevention, but understanding may.

Notwithstanding these criticisms, however, if accidents need to be assigned to categories (perhaps for some actuarial or political reason), then HFACS is probably the method of choice. However, alternatives exist, and the interested reader is directed to Beaubien and Baker's (2002) excellent review of current taxonomic methods applied to aviation accidents for a comprehensive overview and comparative analyses of this topic.

8.5 SPECIAL PROBLEMS IN DOING RESEARCH ON ACCIDENTS

Several problems arise when one is conducting research on the causes of accidents. Arguably, the dominant problem is the scarcity of accidents. Before the reader tears this book to shreds and writes an angry letter to the editor of his or her local newspaper condemning bloodthirsty aviation researchers, let us state that we are not wishing for more accidents. Rather, we are pointing out that the prediction of rare events presents some special difficulties from a statistical standpoint, the details of which are beyond the scope of this book. Some of the difficulties can be illustrated with a simple example, however.

Let us assume that we are interested in predicting the occurrence of fatal accidents among general aviation pilots in the United States. There are roughly 300,000 general aviation pilots in the United States and every year about 300 fatal accidents occur. (For the sake of convenience, let us use rounded numbers that are approximately accurate.) Thus, if accidents are purely random, the chance of any particular pilot being in a fatal accident is 0.001 (300 divided by 300,000) or one out of a thousand. Imagine, then, if we are trying to evaluate whether pilots with beards have more fatal accidents than pilots without beards. We might begin our study in January by identifying a group of 100 pilots, half with beards and half without. Then in December we would see how many are still alive. Fortunately for the pilots, but unfortunately for our research project, all the pilots will almost certainly still be alive at the end of the year—thus telling us nothing about the accident-causing properties of beards. The difficulty is that if only one pilot out of a thousand has an accident during a year, then we would expect something less than one pilot out of the hundred in our study group to have an accident. Unless the effect upon safety of having (or not having) a beard is tremendous, we are unlikely to obtain any results of interest.

Technically, what we are lacking here is variance (or variability) in the criterion (accident involvement). If there is no variance, then the observation can contain no information. This is equivalent to trying to find the smartest person in the class by giving all the students a test. If the test is too easy, then all the students get perfect scores, and we cannot tell from the scores who is the smartest student. (For additional information, consult Nunnally [1978] for his description of the effects of extreme p/q splits on the point biserial correlation.)

This limitation leads to the use of nonparametric statistics (e.g., chi-square), which do not require normal distributions (e.g., Poisson and negative-binomial regression), and, on occasion, very large samples. It also leads to the use of measures other than accident involvement as the criteria for studies.

8.5.1 Is a Close Call Almost the Same as an Accident?

It can be argued that accidents can be thought of as the tip of an iceberg. There are relatively few accidents, but there are many more incidents and hazardous events that did not result in an accident. Sometimes the difference between an incident and an accident is very slim—perhaps only a matter of a few feet in clearing a tree on takeoff. Thus, these incidents may represent instances in which, had the circumstances been only slightly different (perhaps a slightly hotter day, just a little more fuel on board, only a few less pounds of air in the tires), the incident would have become an accident.

It has been suggested (Hunter 1995) that incidents and hazardous events could be considered surrogates for the actual measure of interest—accident involvement. Finding a significant relationship between some measure of interest (e.g., some personality trait) and the surrogate measure of number of hazardous events experienced during the previous year would therefore be an indication that the measure may be related to accident involvement. This approach has been taken in several of the studies described later in this chapter.

The prevalence of incidents and hazardous events has been examined in the United States (Hunter 1995) and in New Zealand (O'Hare and Chalmers 1999). Hunter conducted a nationwide survey of U.S. pilots using a set of questions he termed the "hazardous events scale" (HES). The HES consisted of questions that asked respondents to indicate the number of times they were involved in potentially hazardous events. The questions and the responses for private and commercial pilots are given in Table 8.6. For this same group of respondents, approximately 9% of the private pilots and 17% of the commercial pilots reported having been in an aircraft accident

TABLE 8.6

Reports of Hazardous Events among U.S. Pilots

	Private		Commercial	
	No. of Events[a]	Reporting	No. of Events	Reporting
Low fuel incidents	0	80%	0	66%
	1	16%	1	24%
	2	3%	2	7%
	3	1%	3	2%
	≥4	0%	≥4	2%
On-airport precautionary or forced landing	0	54%	0	41%
	1	23%	1	21%
	2	11%	2	15%
	3	4%	3	7%
	≥4	8%	≥4	17%
Off-airport precautionary or forced landing	0	93%	0	82%
	1	5%	1	10%
	2	1%	2	3%
	3	0%	3	2%
	≥4	0%	≥4	3%
Inadvertent stalls	0	94%	0	90%
	1	5%	1	6%
	2	1%	2	2%
	3	0%	3	0%
	≥4	0%	≥4	1%
Become disoriented (lost)	0	83%	0	83%
	1	14%	1	13%
	2	2%	2	3%
	3	0%	3	1%
	≥4	0%	≥4	0%
Mechanical failures	0	55%	0	33%
	1	27%	1	26%
	2	10%	2	17%
	3	4%	3	9%
	≥4	4%	≥4	16%
Engine quit due to fuel starvation	0	93%	0	84%
	1	6%	1	12%
	2	1%	2	3%
	3	0%	3	1%
	≥4	0%	≥4	1%

TABLE 8.6 (CONTINUED)
Reports of Hazardous Events among U.S. Pilots

	Private		Commercial	
	No. of Events[a]	Reporting	No. of Events	Reporting
Flow VFR into IMC	0	77%	0	78%
	1	15%	1	14%
	2	6%	2	5%
	3	1%	3	2%
	≥4	2%	≥4	2%
Become disoriented (vertigo) while in IMC	0	95%	0	91%
	1	4%	1	7%
	2	1%	2	2%
	3	0%	3	0%
	≥4	0%	≥4	0%
Turn back due to weather	0	29%	0	23%
	1	21%	1	16%
	2	19%	2	18%
	3	13%	3	11%
	≥4	22%	≥4	32%

Source: From Hunter, D. R. 1995. Airman research questionnaire: Methodology and overall results (technical report no. DOT/FAA/AM-95/27), Table 21. Washington, D.C.: Federal Aviation Administration.

[a] Number of times this event has been experienced during a flying career.

at some point in their careers. Clearly, these were nonfatal accidents (because the pilots were alive to respond to the survey), which are about five to six times more prevalent than fatal accidents (i.e., there were about 300 fatal accidents in 2005 out of about 2,000 total accidents).

In a survey of New Zealand pilots, O'Hare and Chalmers (1999) found that encounters with potentially hazardous events were fairly common. Furthermore, the experiences reported by the New Zealand pilots were remarkably similar to those of the U.S. pilots. For example, the proportions of pilots who had entered IMC (slightly under 25%) or inadvertently stalled an airplane (about 10%) were almost identical in both samples. It is interesting that, even given the geographic disparity between the United States and New Zealand, the experiences of pilots in the two countries were so similar. This is a finding to which we will return later in this chapter.

Findings such as these have encouraged researchers to consider the utility of such measures in accident research. Because many more incidents than fatal accidents occur (perhaps as many as two or three orders of magnitude), then using them in research, in lieu of accidents, alleviates to some degree the problems associated with the prediction of rare events. However, these are only surrogates for the true criterion of interest, so these results are only suggestive of possible relationships.

8.5.2 OUT OF THE AIR AND INTO THE LABORATORY

Because of the difficulties associated with conducting naturalistic research—that is, research involving pilots engaged in flights—some researchers have chosen to conduct laboratory-based research. Typically, this research utilizes flight simulators of varying degrees of fidelity and flight profiles designed to expose the pilots to the conditions of interest. Laboratory research has the advantage of tight control over the stimuli (e.g., the aircraft capabilities, the weather experienced). However, because the risk of physical injury and death is never present in these situations, it is always a challenge to defend the generalizability of results from the laboratory to actual flight. Both naturalistic and laboratory research have their advantages and disadvantages and it is incumbent upon the researchers to utilize their capabilities appropriately and to note the limitations of their research in their reports.

8.6 WHY ARE SOME PILOTS SAFER THAN OTHERS?

Having touched upon some of the problems of doing research on aviation accidents, let us now turn to some of the work that has been done to help explain why some pilots are more likely to have accidents than others. For this discussion, we will put aside considerations of whether some aircraft are safer than others (they are), whether some training makes pilots safer (it does), or whether other environmental factors such as geographic variables, air traffic control, and maintenance impact safety (they all do). Instead, we will look solely at the psychological characteristics of the pilots—their attitudes, personality, decision-making skills—and how they influence the likelihood of being in an accident.

8.7 THE DECISION-MAKING COMPONENT OF ACCIDENTS

Earlier in the discussion of various attempts to catalog the causes of accidents, the study by Jensen and Benel (1977) was described as a seminal work in that it specifically identified decision making as contributing to a large proportion of fatal general aviation accidents. That study sparked a great deal of interest in how pilots make decisions that put them at risk of being in an accident, and how decision making might be improved. Indeed, one of the first outcomes from this focus on decision making was a report by Berlin et al. (1982a) in which they described a training program aimed specifically at addressing the decision-making shortcomings identified in the Jensen and Benel study.

Working at Embry-Riddle Aeronautical University under the sponsorship of the U.S. Federal Aviation Administration (FAA), Berlin et al. (1982a) developed a training program and student manual that included the following:

- three subject areas
  - pilot—the pilot's state of health, competency in a given situation, level of fatigue, and other factors affecting performance
  - aircraft—considerations of airworthiness, power plant, and performance criteria such as weight and balance

- environment—weather, airfield altitude and temperature, and outside inputs such as weather briefings or ATC instructions

- six action ways
  - do—no do
    - Do—the pilot did something he or she should not have done.
    - No do—the pilot did not do something he or she should have done.
  - under do—over do
    - Under do—the pilot did not do enough when he or she should have done more.
    - Over do—the pilot did too much when he or she should have done less.
  - early do—late do
    - Early do—the pilot acted too early when he or she should have delayed acting.
    - Late do—the pilot acted too late when he or she should have acted earlier.

- poor judgment behavior chain
  - One poor judgment increases the probability that another poor judgment will follow.
  - The more poor judgments made in sequence, the more probable that others will continue to follow.
  - As the poor judgment chain grows, the alternatives for safe flight decrease.
  - The longer the poor judgment chain becomes, the more probable it is that an accident will occur.

- three mental processes of safe flight
  - automatic reaction
  - problem resolving
  - repeated reviewing

- five hazardous thought patterns (having one or more of these hazardous thought patterns predisposed pilots to acting in ways that placed them at greater risk for accident involvement)
  - antiauthority
  - impulsivity
  - invulnerability
  - macho
  - external control (resignation)

The second volume of this report (Berlin et al. 1982b) contained detailed descriptions and exercises corresponding to the preceding elements. This included an instrument that pilots could use to do a self-assessment of their hazardous thought patterns. An initial, small-scale evaluation of the training manual was conducted using three groups of students at Embry-Riddle Aeronautical University. One of the three groups received the new training program, and the other two groups served as controls. In this evaluation, the experimental group had significantly better performance on written tests and on an observation flight than the two control groups. On the basis of these findings, the authors concluded that the training manual had a positive effect on the decision making of the test subjects.

8.8 AERONAUTICAL DECISION MAKING

Based upon the positive initial results, a series of publications using material taken from the Berlin et al. training manual were produced by the FAA. Each publication was tailored to fit the needs and experiences of a particular segment of the pilot population. The publications included:

- "Aeronautical Decision Making for Helicopter Pilots" (Adams and Thompson 1987);
- "Aeronautical Decision Making for Instructor Pilots" (Buch, Lawton, and Livack 1987);
- "Aeronautical Decision Making for Student and Private Pilots" (Diehl et al. 1987);
- "Aeronautical Decision Making for Instrument Pilots" (Jensen, Adrion, and Lawton 1987);
- "Aeronautical Decision Making for Commercial Pilots" (Jensen and Adrion 1988); and
- "Risk Management for Air Ambulance Helicopter Operators" (Adams 1989).

In addition to these publications, which are rather narrowly aimed at training specific skills in defined groups of pilots, Jensen (1995) has produced a text that covers the common elements across all these publications in greater depth. This book also incorporates considerations of crew resource management (CRM), a concept that developed more or less in parallel with the work on pilot decision making. However, although the decision-making work was typically oriented toward general aviation pilots, the work on CRM grew out of reviews of accidents in the air carrier community.

Following the release of the aeronautical decision-making (ADM) training manual by Berlin et al. (1982a, 1982b) and the initial evaluation of the effectiveness of this training in the United States, similar studies were conducted elsewhere. In Canada, a study that evaluated air cadets was conducted in which the judgment of the cadets was tested during a flight by asking them to perform an unsafe maneuver. Pilots who received the ADM training made correct decisions in 83% of the test situations, compared to 43% of the pilots who did not receive such training (Buch and Diehl 1983, 1984). Another study of pilots attending flight schools in Canada also showed a significant impact for the ADM training (Lester et al. 1986). In that study, 70% of the group that received ADM training chose the correct responses on an observation flight, compared to 60% of the pilots in the control group.

Similar results were noted in a study conducted in Australia (Telfer and Ashman 1986; Telfer 1987, 1989) using students from five flying schools in New South Wales. Even though the samples used were quite small (only 20 total subjects divided among three groups), significant differences in favor of the ADM training were found among the groups.

The results from these evaluations of ADM training and others (Diehl and Lester 1987; Connolly and Blackwell 1987) were reviewed by Diehl (1990), who provided a summary of the results in terms of reductions in pilot error. These results are reproduced in Table 8.7.

Given the consistently significant results demonstrated by ADM training in reducing error among pilots in these studies, it seems clear that the training does have an impact on pilot behavior. However, significant issues have not been addressed by these studies. For example, how long does the effect last following training? Because all the evaluations were conducted immediately following the completion of pilot training or shortly thereafter, the rate at which the training effect decays cannot be determined. This is important because, if the effect only lasts a short time, then the training must be repeated frequently to maintain the effect.

Second, what parts of the training are having an impact? Recall from the list of contents presented earlier that ADM training covers a fairly broad spectrum of topics, ranging from decision heuristics (DECIDE model) to personality traits ("five hazardous attitudes"). Because these were all covered at the same time during the ADM training, it is not possible from the existing data to determine whether they are all needed or whether only one or two of the individual components are responsible for the improvements in behavior.

Finally, an inspection of the data in Table 8.7 suggests that the venue in which the training is administered may influence the magnitude of the effect. Specifically, it is interesting to note that the smallest effects were noted in the least rigorous training environments (the aero clubs and fixed-base operators); much larger effects were found in the aeronautical university and flight school environments.

All of these would be important issues to consider and investigate more fully. Indeed, this approach to developing an intervention without an underlying theoretical rationale and without a firm empirical basis has exposed ADM training to some

TABLE 8.7
Results of ADM Training Evaluation Studies

Researchers	Environment	Experiment n	Control n	Error Reduction
Berlin et al. (1982)	Aeronautical university	26	24	17%
Buch and Diehl (1983)	Flying schools	25	25	40%
Buch and Diehl (1983)	College	17	62	9%
Telfer and Ashman (1986)	Aero clubs	8	6	8%
Diehl and Lester (1987)	Fixed-base operators	20	25	10%
Connolly and Blackwell (1987)	Aeronautical university	16	16	46%

Source: Adapted from Diehl, A. E. 1990. In *Proceedings of the 34th Meeting of the Human Factors Society*, 1367–1371, Table 1. Santa Monica, CA: Human Factors Society.

Note: All results significant, $p < .05$.

criticism (cf. O'Hare and Roscoe 1990; Wiggins and O'Hare 1993). These criticisms are reflected in studies of one major component of ADM training: the five hazardous thoughts or, as they are sometimes called, "hazardous attitudes."

8.9 HAZARDOUS ATTITUDES

All of the ADM training manuals, including the FAA publications listed earlier, have included an instrument for the self-assessment of hazardous attitudes. Although the content varies with the specific group of pilots for whom the training was designed, all the instruments consist of some number (typically 10) of scenarios in which an aviation situation is described. Five alternative explanations for the course of action taken by the pilot in the scenario are then provided and the pilot completing the instrument is asked to choose the one that he or she thinks best applies. The following example is taken from the FAA publication aimed at student and private pilots (Diehl et al. 1987):

- You have just completed your base leg for a landing on runway 14 at an uncontrolled airport. As you turn to final, you see that the wind has changed, blowing from about 90 degrees. You make two sharp turns and land on runway 8. What was your reasoning?
  a. You believe you are a really good pilot who can safely make sudden maneuvers.
  b. You believe your flight instructor was overly cautious when insisting that a pilot must go around rather than make sudden course changes while on final approach.
  c. You know there would be no danger in making the sudden turns because you do things like this all the time.
  d. You know landing into the wind is best, so you act as soon as you can to avoid a crosswind landing.
  e. The unexpected wind change is a bad break, but you figure if the wind can change, so can you.

Each of the five alternatives is keyed to one of the five hazardous attitudes. In this example, the keyed attitudes are (a) macho, (b) antiauthority, (c) invulnerability, (d) impulsivity, and (e) resignation. (Thus, the person who selected alternative *a* as the best explanation for the behavior of the pilot in the scenario would be espousing a macho attitude.) Using the scoring key provided in the training manuals, pilots can compute their scores for each of the five hazardous attitudes and can create a profile of their hazardous attitudes. From that profile, they may identify which, if any, of the attitudes is dominant. The text of the training manuals then provides some guidance on dealing with each of the hazardous attitudes and proposes some short, easily remembered "antidotes" for each attitude. For example, the antidote for having a macho attitude is that "taking chances is foolish" (Diehl et al. 1987, p. 63).

Three studies (Lester and Bombaci 1984; Lester and Connolly 1987; Lubner and Markowitz 1991) have compared the hazardous attitudes scales with other personality measures. These have included the Rotter locus of control scale and several scales from the Cattell 16 PF. In all of these studies, the scores for the individual hazardous attitudes were found to be highly correlated with each other. Moderate to low correlations were also observed with the other personality measures as well as with external criteria such as involvement in near-accidents. However, the interpretation of the results from these studies is problematic.

Hunter (2004) has criticized the use of the self-assessment instrument used in the various FAA publications because it utilized an ipsative format. As Anastasi (1968, p. 453) notes, an ipsative scale is one in which "the strength of each need is expressed, not in absolute terms, but in relation to the strength of the individual's other needs... an individual responds by expressing a preference for one item against another." In this type of scale, having a high score on one subscale forces the scores on the other subscales to be low. Because of this restriction, ipsative scales are subject to significant psychometric limitations. These limitations make the use of traditional statistical analysis methods (such as correlation) inappropriate in many instances. Hence, studies that have tried to correlate hazardous attitude scores taken from the self-assessment instrument with other criteria (such as scores on other psychological instruments) are seriously flawed and cannot, for the most part, provide useful and reliable information. (For a discussion of the difficulties associated with ipsative scoring methods, see Bartram 1996; Saville and Wilson 1991.)

To address this problem, Hunter (2004) recommended that researchers use scales based on Likert scale items. This type of item, widely used in psychological research, typically consists of a statement (e.g., "I like candy") to which the respondent expresses his or her degree of agreement by selecting one of several alternatives (e.g., "strongly agree," "agree," "disagree," or "strongly disagree").

Hunter and other researchers (Holt et al. 1991) have developed instruments for the assessment of hazardous attitudes using Likert scale items. In a comparison of the traditional instrument contained in the FAA training materials and the Likert-style instruments (Hunter 2004), the superiority of the Likert instruments in terms of reliability and in correlations with external criteria (involvement in hazardous events) was clearly demonstrated. Using these measurement instruments, it is possible to demonstrate empirically that pilots' attitudes can affect the likelihood of their involvement in an accident.

Hazardous attitudes are but one of several psychological constructs that have been considered as possible factors that impact the likelihood of accident involvement. These constructs include locus of control, risk perception, risk tolerance, and situational awareness.

8.10 LOCUS OF CONTROL

Locus of control (LOC) refers to the degree to which a person believes that what happens to him or her is under his or her personal control (internal LOC) or that what happens is a result of external factors (e.g., luck, the actions of others) over which he or she has no control (external LOC). This construct was first proposed by Rotter [1966] and since that time it has been used in a variety of settings (see Stewart [2006] for a review). Wichman and Ball (1983) administered the LOC to a sample of 200 general aviation pilots and found that the pilots were significantly more internal than Rotter's original sample. They also found that the pilots who were higher on LOC internality were more likely to attend safety clinics, possibly indicating a greater safety orientation among this group compared to the pilots with a greater LOC externality.

Several variations on the Rotter LOC scale have been constructed that assess LOC perceptions in a particular domain. These include LOC scales that are specific to driving (Montag and Comrey 1987) and medical issues (Wallston et al. 1976). These development efforts were spurred by the belief, noted by Montag and Comrey (1987, p. 339), that "attempts to relate internality–externality to outside criteria have been more successful when the measures of this construct were tailored more specifically to the target behavior (e.g., drinking, health, affiliation), rather than using the more general I–E scale itself."

Continuing in this same vein, Jones and Wuebker (1985) developed and validated a safety LOC scale to predict employees' accidents and injuries in industrial settings. They found that participants in the lower accident risk groups were significantly more internal on the safety LOC than participants in the high-risk groups. In a subsequent study of safety among hospital workers, Jones and Wuebker (1993) found that workers who held more internal safety attitudes were significantly less likely to have an occupational accident compared to employees with more external attitudes.

Based upon these results, Hunter (2002) developed an aviation safety locus of control (AS-LOC) scale by modifying the Jones and Wuebker (1985) scale so as to put all the scale items into an aviation context. Two example items from the AS-LOC measuring internality and externality, respectively, are

- Accidents and injuries occur because pilots do not take enough interest in safety.
- Avoiding accidents is a matter of luck.

In an evaluation study conducted using 176 pilots who completed the AS-LOC over the Internet, Hunter (2002) found a significant correlation ($r = -.205$; $p < .007$) between internality and involvement in hazardous events (specifically, the hazardous event scale described earlier). In contrast, a nonsignificant correlation ($r = .077$) was found between externality and the hazardous event scale score. Consistent with the previous research, pilots exhibited a substantially higher internal orientation than external orientation on the new scale.

Similar findings were also reported by Joseph and Ganesh (2006), who administered the AS-LOC to a sample of 101 Indian pilots. As in the previous research, the Indian pilots also had significantly higher internal than external LOC scores. An interesting finding of this study is that the civil pilots had higher internal LOC scores than the military pilots. Additionally, the transport pilots had the highest internal scores of any of the pilot groups, followed by fighter pilots and helicopter pilots. Given the differences in accident rates among these groups, it would be interesting to investigate the degree to which these differences in AS-LOC scores are attributable to training, formal selection processes, or some sort of self-selection process.

8.11   RISK PERCEPTION AND RISK TOLERANCE

Risk assessment and management make up one component of the broader process of pilot decision making. As noted earlier, poor pilot decision making has been implicated as a leading factor in fatal general aviation accidents (Jensen and Benel 1977), and poor risk assessment can contribute significantly to poor decision making. To address the question of risk perception among pilots, O'Hare (1990) developed an aeronautical risk judgment questionnaire to assess pilots' perceptions of the risks and hazards of general aviation. Hazard awareness was assessed by having pilots

1. estimate the percentage of accidents attributable to six broad categories;
2. rank the phases of flight by hazard level; and
3. rank detailed causes of fatal accidents (e.g., spatial disorientation, misuse of flaps).

O'Hare found that pilots substantially underestimated the risk of general aviation flying relative to other activities and similarly underestimated their likelihood of being in an accident. Based on these results, he concluded that "an unrealistic assessment of the risks involved may be a factor in leading pilots to 'press on' into deteriorating weather" (O'Hare 1990, p. 599).

This conclusion was supported by research (O'Hare and Smitheram 1995; Goh and Wiegmann 2001) that shows that pilots who continue flight into adverse weather conditions have a poor perception of the risks. Interestingly, similar results are found in studies of youthful drivers (Trankle, Gelau, and Metker 1990), who have significantly poorer perceptions of the hazards involved in driving compared to older, safer drivers.

Risk perception and risk tolerance are related and often confounded constructs. Hunter defined risk perception as "the recognition of the risk inherent in a situation" (2002, p. 3) and suggested that risk perception may be mediated both by the characteristics of the situation and the characteristics of the pilot experiencing the situation. Therefore, situations that present a high level of risk for one person may present only low risk for another. For example, the presence of clouds and low visibility may present a very high risk for a pilot qualified to fly only under visual meteorological conditions (VMC), but the same conditions would present very little risk for an experienced pilot qualified to fly in instrument meteorological conditions (IMC) in an appropriately equipped aircraft.

The pilot must therefore accurately perceive not only the external situation, but also his or her personal capacities. Underestimation of the external situation and overestimation of personal capacity lead to a misperception of the risk and are frequently seen as a factor in aircraft accidents. Risk perception may therefore be conceived as primarily a cognitive activity involving the accurate appraisal of external and internal states.

By contrast, Hunter defined risk tolerance as "the amount of risk that an individual is willing to accept in the pursuit of some goal" (2002, p. 3). Risk tolerance may be affected by the person's general tendency to risk aversion as well as the personal value attached to the goal of a particular situation. In flying, just as in everyday life, some goals are more important than others; the more important the goal is, the more risk a person may be willing to accept.

Noting that previous studies had assessed pilots' estimates of global risk levels for broad categories (e.g., pilot, weather, etc.) and drawing upon the extensive driver research, Hunter (2002, 2006) proposed that more specific measures of risk perception and risk tolerance were needed. These new measures would operate at a tactical level involving specific aviation situations, as opposed to the strategic level measures used previously. Using this approach, Hunter developed two measures of risk perception and three measures of risk tolerance.

The risk perception measures included one measure (risk perception—self), that asked pilots about the risk they personally would experience in a set of situations and another measure (risk perception—other) that asked pilots about the risk that some other pilot would experience in another set of situations. Examples of both types of measures follow:

- risk perception—other
  - Low ceilings obscure the tops of the mountains, but the pilot thinks that he can see through the pass to clear sky on the other side of the mountain ridges. He starts up the wide valley that gradually gets narrower. As he approaches the pass he notices that he occasionally loses sight of the blue sky on the other side. He drops down closer to the road leading through the pass and presses on. As he goes through the pass, the ceiling continues to drop and he finds himself suddenly in the clouds. He holds his heading and altitude and hopes for the best.
  - The pilot is in a hurry to get going and does not carefully check his seat, seat belt, and shoulder harness. When he rotates, the seat moves backward on its tracks. As it slides backward, the pilot pulls back on the control yoke, sending the nose of the aircraft upward. As the airspeed begins to decay, he strains forward to push the yoke back to a neutral position.
  - Just after takeoff, a pilot hears a banging noise on the passenger side of the aircraft. He looks over at the passenger seat and finds that he cannot locate one end of the seatbelt. He trims the aircraft for level flight, releases the controls, and tries to open the door to retrieve the seatbelt.

- risk perception—self
  - At night, fly from your local airport to another airport about 150 miles away, in a well-maintained aircraft, when the weather is marginal VFR (3 miles visibility and 2,000 feet overcast).
  - Fly in clear air at 6,500 feet between two thunderstorms about 25 miles apart.
  - Make a traffic pattern so that you end up turning for final with about a 45° bank.

In the case of the risk perception—other scale, pilots were asked to rate the risk for a third-party, low-time general aviation pilot, using a scale from 1 (very low risk) to 100 (very high risk). For the risk perception—self scale, pilots were asked to rate the risk if they personally were to perform this tomorrow (also using the 1–100 rating scale).

To measure risk tolerance, Hunter (2002, 2006) created three variants of a risky gamble scenario, each in an aviation setting. One variant involved making repeated flights in aircraft with a known likelihood of mechanical failure; the other two involved flights between thunderstorms of varying distance and through mountainous areas with deteriorating weather. All three variants were structured so that the participants could gain points (and potential prizes) by accepting the risk and taking a flight; however, if they failed to complete a flight (i.e., crashed), they would lose points. This manipulation was intended to provide a motivation to complete flights while at the same time encouraging some degree of caution because crashes could result in the loss of real prizes.

These measures and several others were administered to a large sample of pilots over the Internet. In general, support was found for the risk perception scales in terms of their correlations with pilot involvement in hazardous aviation events. Pilots who experienced more hazardous events tended not to have rated the scenarios as risky, compared to pilots with fewer hazardous events. However, the measures of risk tolerance were not significantly correlated with hazardous aviation events. This led Hunter to conclude that poor perception of risks was a more important predictor of hazardous aviation events and, by extension, of aviation accident than was risk tolerance.

8.12 SITUATION AWARENESS

In common terms, situation awareness (SA) means knowing what is going on around one. For a pilot, this means knowing where the aircraft is with respect to other aircraft in the vicinity, important objects on the ground (e.g., runways, mountains, tall radio towers), and weather elements such as clouds, rain, and areas of turbulence. In addition, SA means knowing what the aircraft is doing at all times, both externally (e.g., turning, descending) and internally (e.g., fuel status, oil pressure). (Students of Eastern philosophy may recognize this Zen-like state as mindfulness, or being "one with the moment.") Moreover, SA has both a present and a future component. Thus, having good SA means that a pilot knows all the things going on "right now" and can reliably estimate what will be happening a few minutes or a few hours from now. This distinction is reflected in the Endsley's proposed definition of SA as "the perception of the elements in the environment within a volume of time and space, the comprehension of their meaning and the projection of their status in the near future" (1988, p. 87).

Situation awareness is particularly important in the framework of Klein's recognition-primed decision (RPD) model that emphasizes the importance of SA (Kaempf et al. 1996; Klein 1993). Klein's RPD model suggests that pilots actually perform little real problem solving. Rather, the major activity is recognizing a situation and then selecting one of the limited number of solutions that have worked in the past. Clearly, in such a model, awareness of surroundings is very important for the detection of changes in the environment that may be used as part of the recognition process.

Within the RPD theoretical framework, experience is critical because it builds the repertoire by which one may accurately identify the salient cues and correctly diagnose the situation. According to Klein (2000, p. 174),

> The most common reason for poor decisions is a lack of experience. It takes a high degree of experience to recognize situations as typical. It takes a high degree of experience to build stories to diagnose problems and to mentally simulate a course of action. It takes a high degree of experience to prioritize cues, so workload won't get too high. It takes a high degree of experience to develop expectancies and to identify plausible goals in a situation.

Although SA is an intriguing construct, it could be argued that it is simply a meta-construct, incorporating some or all of the other, more basic constructs previously discussed. For example, from Endsley's (1988) definition, SA would subsume the risk perception elements discussed earlier because proper detection and evaluation of the cues associated with, for example, deteriorating weather conditions would fall within the definition of SA. The same argument could be made regarding self-knowledge of internal states such as attitudes. The question, therefore, is whether SA is something more than the sum of the constituent parts. Is it simply another category to which behavior and accidents may be consigned, without delving into an understanding of why they occur? As noted earlier, describing is not the same thing as explaining. The present authors suggest that the latter description is more accurate, but the interested reader may wish to consult the literature. (The book by Endsley and Garland [2000] is a good source.)

8.13   AVIATION WEATHER ENCOUNTERS

Over the years, encounters with adverse weather have remained one of the largest single causes of fatal general aviation accidents. Particularly interesting are those instances in which the pilot continued a flight from visual to instrument conditions and subsequently lost control of the aircraft or struck the ground while trying to exit the weather. Several researchers have examined these accidents from a variety of perspectives. One such perspective is to focus on these events as a plan continuation error (Orasanu, Martin, and Davison 2001). This perspective suggests that pilots fail to alter their plans when unforeseen conditions are encountered that make the original plan untenable. This failure can be attributed to the risk perception and risk tolerance constructs suggested by Hunter (2006).

It can also be interpreted in terms of sunk costs (O'Hare and Smitheram 1995). The sunk-cost concept attempts to explain plan continuation error as arising from the inherent desire of the pilot not to waste previous efforts. That is, once a trip has been initiated, each minute of the trip represents the expenditure of resources (time and money) that will be lost if the pilot is forced to return to the origination point without completing the flight. Early in a flight, this potential loss is relatively small, but as the duration of the flight grows, so does the potential loss. This potential loss (the "sunk cost" in accounting terms) then represents a motivation to continue the flight, even into marginal conditions.

O'Hare and Owen (1999) tested this concept by having pilots fly a simulated cross-country flight in which they encountered adverse weather either early or late in the flight. The sunk costs concept would predict that the pilots who encountered the weather later in the flight would be more likely to press on in an attempt to reach their destination. However, in this experiment, the results failed to support that hypothesis: A majority of pilots in both conditions diverted their flights. Thus, the validity of this concept as an explanation for pilot behavior in the face of adverse weather is questionable.

In a different approach to understanding pilot behavior with respect to weather, Hunter, Martinussen, and Wiggins (2003) used a mathematical modeling technique to examine the manner in which pilots combined information about visibility, cloud ceiling, precipitation, and terrain to make judgments about the safety of a flight. In this study, pilots in the United States, Norway, and Australia were given three maps depicting flights in their respective countries. One map depicted a flight over level terrain, and another showed a flight over mountainous terrain. The final map depicted a flight over a large body of water.

A scenario-based judgment task in which a safety rating was provided for each of 27 weather scenarios for each of the three routes was then completed by 326 American, 104 Norwegian, and 51 Australian pilots. The 27 weather scenarios were based on combinations of varying levels of visibility, ceiling, and precipitation. These safety ratings were then used to develop individual regression equations for each pilot. (For a discussion of regression equations, see Chapter 2 on statistics.) The regression equation for a pilot described the information combination process that he or she used to assign the safety ratings.

Two interesting results were observed. First, the safety ratings for the 27 scenarios were very similar for the three diverse groups of pilots. Second, for each group, the compensatory model of information use was favored over noncompensatory models. The use of a compensatory weather model means that a pilot might decide that conditions are suitable for flight when the ceiling is high (a safe situation) but the visibility is low (an unsafe situation) because the high ceiling compensates for the low visibility in the overall evaluation of the situation.

In contrast, in a typical noncompensatory model (referred to as the multiple-hurdle model), each aspect of the situation is individually examined and compared to a criterion. A decision to initiate a flight is made only if all the factors individually meet their respective criteria. Here, a high value on one variable cannot compensate for a low value on another variable. Hunter et al. argue that using a compensatory decision model puts inexperienced pilots at greater risk of being in an accident.

Overconfidence by pilots was investigated by Goh and Wiegmann (2001), who found that pilots who continued into weather conditions in a simulated flight reported greater confidence in their piloting abilities, even though there were no differences in training or experience when compared to pilots who chose to divert. These same pilots also judged weather and pilot error as less likely threats to flight safety than the pilots who diverted and they believed themselves less vulnerable to pilot error.

8.14   OTHER PROGRAMS TO IMPROVE SAFETY

Under the sponsorship of the FAA, researchers at Ohio State University began a program of research in the early 1990s aimed at developing better understanding of the causes of accidents among general aviation pilots, with the explicit goal of developing interventions to improve safety. The approach of this research was focused more on the development of expertise among relatively inexperienced pilots than on assessing hazardous thoughts or providing heuristics for decision making (Kochan et al. 1997). This work led to the development of three training products aimed at improving decision making by pilots (1) during the preflight planning process, (2) when making decisions in flight, and (3) when making weather-related decisions.

The first of these three products trained pilots to recognize the hazards present in flights and to establish a set of minimum operating standards (usually termed "personal minimums") that would create a buffer against those hazards (Kirkbride et al. 1996). For example, although it is legal to fly at night in the United States with 4 miles' visibility and a ceiling of 2,000 feet, a prudent pilot lacking an instrument rating might elect to fly at night only when the visibility is greater than 8 miles and the ceiling is over 5,000 feet. These more stringent standards become that pilot's personal minimums and are recorded in a personal checklist that pilots are encouraged to review before each flight. Evaluations of pilot acceptance of this new training were positive (Jensen, Guilkey, and Hunter 1998), although no evaluation was conducted of the impact of the training on external criteria such as involvement in hazardous events or accidents.

In contrast to the attempt at procedural standardization incorporated in the personal minimums training and the hazardous thoughts training contained in the several FAA publications, a skills-based approach has been proposed that would focus on helping pilots improve their skill at recognizing and dealing with hazardous situations. O'Hare and colleagues (1998) utilized the techniques of cognitive task analysis (CTA) and the critical decision method (CDM) form of CTA described by Klein, Calderwood, and MacGregor (1989) to evaluate the decision processes of highly experienced general aviation pilots in adverse weather situations. Use of this technique allowed them to identify the information cues and processes used by these expert pilots in making weather-related decisions. Using these data, Wiggins and O'Hare (1993, 2003a), under contract to the FAA, constructed a training program they called WeatherWise.

WeatherWise is a computer-based training program "designed to provide visual pilots with the skills necessary to recognize and respond to the cues associated with deteriorating weather conditions during flight" (Wiggins and O'Hare 2003b, p. 337). The program consists of four stages:

- Stage 1. An assessment is made of in-flight weather conditions from still images to demonstrate the difficulty in making determinations of visual flight conditions.
- Stage 2. An introduction is given to the salient weather cues identified in previous research as being used by experts to make weather decisions. These cues were

- cloud base
- visibility
- cloud coloring
- cloud density
- terrain clearance
- rain
- horizon
- cloud type
- wind direction
- wind speed
- Stage 3. A number of images of in-flight weather conditions are presented to identify the point at which a significant deterioration had taken place. During this stage, a rule of thumb was advocated to the effect that, if a significant deterioration occurred in three or more of the cues, a weather-related decision (possibly a diversion) should be made.
- Stage 4. Further practice in attending to the salient weather cues is undertaken. In this stage, participants view a sequence of in-flight video recordings and identify the point at which conditions deteriorated below visual flight requirements.

This training program was evaluated using a group of 66 Australian private pilots, none of whom had more than 150 total flight hours. In comparison to the control group, who did not receive the training, the pilots who completed the WeatherWise training were significantly more likely to make a diversion decision at or before the optimal point. In contrast, the pilots who did not receive the training tended to continue on into the adverse weather conditions (Wiggins and O'Hare 2003b).

In addition to the FAA, other civil aviation authorities have recognized the need to improve general aviation safety and have incorporated the previously mentioned training programs as part of their national safety efforts. The civil aviation authorities of Australia and New Zealand have adopted the personal minimums and WeatherWise training programs and have distributed the training to their general aviation pilots.

In recognition of the importance of decision making to accident involvement, the FAA, in cooperation with a coalition of aviation industry organizations, formed a Joint Safety Analysis Team (JSAT) to examine general aviation ADM and to develop a program to improve ADM so as to reduce the number of accidents attributable to poor decision making. The JSAT, in turn, chartered an international panel of human factors experts to address the technical issues of how poor decision making contributed to accidents and what might be done to improve aviation safety. That panel's recommendations, listing over 100 specific items, was adopted without change by the JSAT and provided to the FAA as part of its final report (Jensen et al. 2003). Reflecting a pragmatic approach to applying the current knowledge of accident causality among general aviation pilots, the panel's recommendations covered a wide range of possible interventions. Some examples include:

- Create and disseminate to pilots a weather hazard index that incorporates the weather risks into a single graphic or number.
- Reorganize weather briefings so as to present information related to potentially hazardous conditions as the first and last items given to the pilot.
- Increase the use of scenario-based questions in the written examination.
- Include training for certified flight instructors (CFIs) on risk assessment and management in instructional operations.
- Produce a personal minimums checklist training program expressly for use by CFIs in setting their instructional practices.
- Establish a separate weather briefing and counseling line for low-time pilots.
- Require pitot heat to be applied automatically, whenever the aircraft is in flight.
- Develop displays that depict critical operational variables in lieu of raw, unprocessed data (e.g., have fuel indicators that show remaining range or endurance, as well as remaining gallons of fuel).
- Develop and disseminate training that explicitly addresses the issues involved in crash survivability, including crash technique, minimizing vertical loads, and planning for crashes (water, cell phone, matches, etc.) even on flights over hospitable terrain.
- Develop role-playing simulations in which pilots can observe modeled methods of resisting social pressures and can then practice the methods.

Regrettably, these interventions have not yet been implemented, even though they were accepted by both industry and government regulators. This is a reflection, perhaps, of the difficulty of making even well-regarded changes in an established bureaucracy and cost-conscious industry. Clearly, it is not enough for researchers to find better ways to keep pilots safe. They must also find ways to get their discoveries implemented—arguably, the more difficult of the two tasks.

Nevertheless, some progress is being made in training pilots to be more safety conscious. In 2006, the AOPA Air Safety Foundation (ASF) began sending a free DVD on decision making to all newly rated private and instrument pilots. The scenarios contained on the DVD focus on VFR into instrument conditions and IFR decision making—two areas that the ASF has found to be particularly troublesome (Aircraft Owners and Pilots Association 2006).

8.15 SUMMARY

In this chapter we have examined the issue of safety from the perspective of aviation psychology. We have seen that although flying in large commercial air carriers is quite safe, the situation is not so comforting in general aviation, where the risks of involvement in a fatal aviation accident are somewhat higher than being involved in a fatal motor vehicle accident. Curiously, anecdotal evidence (from the responses of many general aviation pilots when this topic is raised at flight safety seminars) suggests that general aviation pilots are largely unaware of this differential risk and generally believe that they are safer when flying than when driving their cars. Hence, programs to improve safety often receive little more than lip service because the pilots involved do not really feel that they are at risk.

Scientific research has identified several factors that place pilots at greater risk of accident involvement. Among those discussed earlier were feelings of invulnerability (hazardous attitudes) and feelings of being a victim of outside forces (locus of control), along with issues relating to recognition of the risks inherent to flight. From this research, programs have been developed to make pilots aware of these risk factors and to train them to recognize the cues that indicate situations of heightened risk requiring immediate action on their part.

The advanced technology formerly found only in air carriers and executive jets is now working its way into the general aviation fleet. This technology will make some tasks easier (e.g., navigation); however, it will present its own set of unique problems and will still require pilots to make reasoned judgments about when, where, how, and if they should undertake a flight. The influence of pilots' personalities and their skill at acquiring and using information will still be great, even in the aircraft of tomorrow. Safety requires a proactive approach to assessing and managing all the elements that influence the outcome of a flight, including the most important element—the human at the controls.

RECOMMENDED READING

Dekker, S. 2006. *The field guide to understanding human error.* Aldershot, England: Ashgate.
Perrow, C. 1984. *Normal accidents: Living with high-risk technologies.* New York: Basic Books.
Reason, J. 1990. *Human error.* Cambridge: Cambridge University Press.
———. 1997. *Managing the risks of organizational accidents.* Aldershot, England: Ashgate.

REFERENCES

Adams, R. J. 1989. Risk management for air ambulance helicopter operators (technical report DOT/FAA/DS-88/7). Washington, D.C.: Federal Aviation Administration.
Adams, R. J., and Thompson, J. L. 1987. Aeronautical decision making for helicopter pilots (technical report DOT/FAA/PM-86/45). Washington, D.C.: Federal Aviation Administration.
Anastasi, A. 1968. *Psychological testing.* New York: Macmillan.
AOPA (Aircraft Owners and Pilots Association). 2006. *The Nall report.* Frederick, MD: Author.
ATSB (Australian Transport Safety Bureau). 2007. Data and statistics. http://www.atsb.gov.au/aviation/statistics.aspx (accessed June 7, 2007).
Bartram, D. 1996. The relationship between ipsatized and normative measures of personality. *Journal of Occupational and Organizational Psychology* 69:25–39.
Beaubien, J. M., and Baker, D. P. 2002. A review of selected aviation human factors taxonomies, accident/incident reporting systems, and data reporting tools. *International Journal of Applied Aviation Studies* 2:11–36.
Berlin, J. I., Gruber, E. V., Holmes, C. W., Jensen, R. K., Lau, J. R., and Mills, J. W. 1982a. Pilot judgment training and evaluation, vol. I (technical report DOT/FAA/CT-81/56-I). Washington, D.C.: Federal Aviation Administration.
———. 1982b. Pilot judgment training and evaluation, vol. II (technical report DOT/FAA/CT-81/56-II). Washington, D.C.: Federal Aviation Administration.
Buch, G. D., and Diehl, A. E. 1983. Pilot judgment training manual validation. Unpublished Transport Canada report. Ontario.

———. 1984. An investigation of the effectiveness of pilot judgment training. *Human Factors* 26:557–564.

Buch, G., Lawton, R. S., and Livack, G. S. 1987. Aeronautical decision making for instructor pilots (technical report DOT/FAA/PM-86/44). Washington, D.C.: Federal Aviation Administration.

Connolly, T. J., and Blackwell, B. B. 1987. A simulator-based approach to training in aeronautical decision making. In *Proceedings of the Fourth International Symposium of Aviation Psychology*. Columbus: Ohio State University.

Dekker, S. W. A. 2001. The disembodiment of data in the analysis of human factors accidents. *Human Factors and Aerospace Safety* 1:39–57.

Diehl, A. E. 1989. Human performance aspects of aircraft accidents. In *Aviation psychology*, ed. R. S. Jensen, 378–403. Brookfield, VT: Gower Technical.

———. 1990. The effectiveness of aeronautical decision making training. In *Proceedings of the 34th Meeting of the Human Factors Society*, 1367–1371. Santa Monica, CA: Human Factors Society.

———. 1991. Does cockpit management training reduce aircrew error? Paper presented at the 22nd International Seminar of the International Society of Air Safety Investigators. Canberra: November 1991.

Diehl, A. E., and Lester, L. F. 1987. Private pilot judgment training in flight school settings (technical report DOT/FAA/AM-87/6). Washington, D.C.: Federal Aviation Administration.

Diehl, A. E., Hwoschinsky, P. V., Lawton, R. S., and Livack, G. S. 1987. Aeronautical decision making for student and private pilots (technical report DOT/FAA/PM-86/41). Washington, D.C.: Federal Aviation Administration.

Endsley, M. R. 1988. Design and evaluation for situation awareness enhancement. In *Proceedings of the Human Factors Society 32nd Annual Meeting* 1:97–101. Santa Monica, CA: Human Factors Society.

Endsley, M. R., and Garland, D. J. 2000. *Situation awareness analysis and measurement*. Mahwah, NJ: Lawrence Erlbaum Associates.

Federal Aviation Administration. 1991. Aeronautical decision making (advisory circular 60-22). Washington, D.C.: FAA.

———. 2007. Administrator's fact book. http://www.faa.gov/about/office_org/headquarters_offices/aba/admin_factbook/ (accessed May 15, 2007).

Gaur, D. 2005. Human factors analysis and classification system applied to civil aircraft accidents in India. *Aviation, Space and Environmental Medicine* 76:501–505.

Goh, J., and Wiegmann, D. A. 2001. Visual flight rules flight into instrument meteorological conditions: An empirical investigation of the possible causes. *International Journal of Aviation Psychology* 11:359–379.

Holt, R. W., Boehm-Davis, D. A., Fitzgerald, K. A., Matyuf, M. M., Baughman, W. A., and Littman, D. C. 1991. Behavioral validation of a hazardous thought pattern instrument. In *Proceedings of the Human Factors Society 35th Annual Meeting*, 77–81. Santa Monica, CA: Human Factors Society.

Hunter, D. R. 1995. Airman research questionnaire: Methodology and overall results (technical report no. DOT/FAA/AM-95/27). Washington, D.C.: Federal Aviation Administration.

———. 2002. Development of an aviation safety locus of control scale. *Aviation, Space, and Environmental Medicine* 73:1184–1188.

———. 2004. Measurement of hazardous attitudes among pilots. *International Journal of Aviation Psychology* 15:23–43.

———. 2006. Risk perception among general aviation pilots. *International Journal of Aviation Psychology* 16:135–144.

Hunter, D. R., Martinussen, M., and Wiggins, M. 2003. Understanding how pilots make weather-related decisions. *International Journal of Aviation Psychology* 13:73–87.

Inglis, M., Sutton, J., and McRandle, B. 2007. Human factors analysis of Australian aviation accidents and comparison with the United States (aviation research and analysis report—B2004/0321). Canberra: Australia Transport Safety Bureau.

Jensen, R. S. 1995. *Pilot judgment and crew resource management.* Brookfield, VT: Ashgate.

Jensen, R. S., and Adrion, J. 1988. Aeronautical decision making for commercial pilots (technical report DOT/FAA/PM-86/42). Washington, D.C.: Federal Aviation Administration.

Jensen, R. S., Adrion, J., and Lawton, R.S. 1987. Aeronautical decision making for instrument pilots (technical report DOT/FAA/PM-86/43). Washington, D.C.: Federal Aviation Administration.

Jensen, R. S., and Benel, R. A. 1977. Judgment evaluation and instruction in civil pilot training (technical report FAA-RD-78-24). Washington, D.C.: Federal Aviation Administration.

Jensen, R. S., Guilkey, J. E., and Hunter, D. R. 1998. An evaluation of pilot acceptance of the personal minimums training program for risk management (technical report DOT/FAA/AM-98/7). Washington, D.C.: Federal Aviation Administration.

Jensen, R. S., Wiggins, M., Martinussen, M., O'Hare, D., Hunter, D. R., Mauro, R., and Wiegmann, D. 2003. Identifying ADM safety initiatives: Report of the human factors expert panel. In *Proceedings of the 12th International Symposium on Aviation Psychology,* ed. R. S. Jensen, 613–618. Dayton: Ohio State University Press.

Jones, J. W., and Wuebker, L. 1985. Development and validation of the safety locus of control scale. *Perceptual and Motor Skills* 61:151–161.

———. 1993. Safety locus of control and employees' accidents. *Journal of Business and Psychology* 7:449–457.

Joseph, C., and Ganesh, A. 2006. Aviation safety locus of control in Indian aviators. *Indian Journal of Aerospace Medicine* 50:14–21.

Kaempf, G., Klein, G., Thordsen, M., and Wolf, S. 1996. Decision making in complex naval command-and-control environments. *Human Factors* 38:220–231.

Kirkbride, L. A., Jensen, R. S., Chubb, G. P., and Hunter, D. R. 1996. Developing the personal minimums tool for managing risk during preflight go/no-go decisions (technical report DOT/FAA/AM-96/19). Washington, D.C.: Federal Aviation Administration.

Klein, G. 2000. *Sources of power: How people make decisions.* Cambridge, MA: MIT Press.

———. 1993. A recognition-primed decision (RPD) model of rapid decision making. In *Decision making in action: Models and methods,* ed. G. Klein, J. Orasanu, R. Calderwood, and C. Zsambok, 138–147. Norwood, NJ: Ablex.

Klein, G., Calderwood, R., and MacGregor, D. 1989. Critical decision method for eliciting knowledge. *IEEE Transactions on Systems, Man, and Cybernetics* 19:462–472.

Kochan, J. A., Jensen, R. S., Chubb, G. P., and Hunter, D. R. 1997. A new approach to aeronautical decision-making: The expertise method (technical report DOT/FAA/AM-97/6). Washington, D.C.: Federal Aviation Administration.

Lester, L. F., and Bombaci, D. H. 1984. The relationship between personality and irrational judgment in civil pilots. *Human Factors* 26:565–572.

Lester, L. F., and Connolly, T. J. 1987. The measurement of hazardous thought patterns and their relationship to pilot personality. In *Proceedings of the Fourth International Symposium on Aviation Psychology,* ed. R. S. Jensen, 286–292. Columbus: Ohio State University.

Lester, L. F., Diehl. A., Harvey, D. P., Buch, G., and Lawton, R. S. 1986. Improving risk assessment and decision making in general aviation pilots. Paper presented at the 57th Annual Meeting of the Eastern Psychological Association. Atlantic City, NJ.

Li, W.-C., and Harris, D. 2005. HFACS analysis of ROC air force aviation accidents: Reliability analysis and cross-cultural comparison. *International Journal of Applied Aviation Studies* 5:65–81.

Lubner, M. E., and Markowitz, J. S. 1991. Rates and risk factors for accidents and incidents versus violations for U.S. airmen. *International Journal of Aviation Psychology* 1:231–243.

Markou, I., Papadopoulos, I., Pouliezos, N., and Poulimenakos, S. 2006. Air accidents–incidents human factors analysis: The Greek experience 1983–2003. Paper presented at the 18th Annual European Aviation Safety Seminar. Athens, Greece.

Montag, I., and Comrey, A. L. 1987. Internality and externality as correlates of involvement in fatal driving accidents. *Journal of Applied Psychology* 72:339–343.

Nunnally, J. C. 1978. *Psychometric theory.* New York: McGraw–Hill.

O'Hare, D. 1990. Pilots' perception of risks and hazards in general aviation. *Aviation, Space, and Environmental Medicine* 61:599–603.

O'Hare, D., and Chalmers, D. 1999. The incidence of incidents: A nationwide study of flight experience and exposure to accidents and incidents. *International Journal of Aviation Psychology* 9:1–18.

O'Hare, D., and Owen, D. 1999. Continued VFR into IMC: An empirical investigation of the possible causes. Final report on preliminary study. Unpublished manuscript, University of Otago, Dunedin, New Zealand.

O'Hare, D., and Roscoe, S. 1990. *Flightdeck performance: The human factor.* Ames: Iowa State University Press.

O'Hare, D., and Smitheram, T. 1995. "Pressing on" into deteriorating conditions: An application of behavioral decision theory to pilot decision making. *International Journal of Aviation Psychology* 5:351–370.

O'Hare, D., Wiggins, M., Williams, A., and Wong, W. 1998. Cognitive task analyses for decision centered design and training. *Ergonomics* 41:1698–1718.

Orasanu, J., Martin, L., and Davison, J. 2001. Cognitive and contextual factors in aviation accidents. In *Linking expertise and naturalistic decision making,* ed. E. Salas and G. Klein, 209–226. Mahwah, NJ: Lawrence Erlbaum Associates.

Perrow, C. 1984. *Normal accidents: Living with high-risk technologies.* New York: Basic Books.

Reason, J. 1990. *Human error.* New York: Cambridge University Press.

———. 1997. *Managing the risks of organizational accidents.* Aldershot, England: Ashgate.

Rotter, J. B. 1966. Generalized expectancies for internal versus external control of reinforcement. *Psychological Monographs* 80 (609): entire issue.

Saville, P., and Wilson, E. 1991. The reliability and validity of normative and ipsative approaches in the measurement of personality. *Journal of Occupational Psychology* 64:219–238.

Shappell, S. A., and Wiegmann, D. A. 2000. The human factors analysis and classification system-HFACS (technical report DOT/FAA/AM-00/7). Washington, D.C.: Federal Aviation Administration.

———. 2002. HFACS analysis of general aviation data 1990–98: Implications for training and safety. *Aviation, Space, and Environmental Medicine* 73:297.

———. 2003. A human error analysis of general aviation controlled flight into terrain (CFIT) accidents occurring between 1990 and 1998 (technical report DOT/FAA/AM-03/4). Washington, D.C.: Federal Aviation Administration.

Shappell, S. A., Detwiler, C. A., Holcomb, K. A., Hackworth, C. A., Boquet, A. J., and Wiegmann, D. A. 2006. Human error and commercial aviation accidents: A comprehensive fine-grained analysis using HFACS (technical report DOT/FAA/AM-06/18). Washington, D.C.: Federal Aviation Administration.

Stewart, J. E. 2006. Locus of control, attribution theory, and the "five deadly sins" of aviation. (technical report 1182). Fort Rucker, AL: U.S. Army Research Institute for the Behavioral and Social Sciences.

Telfer, R. 1987. Pilot judgment training: The Australian study. In *Proceedings of the Fourth International Symposium on Aviation Psychology,* ed. R. S. Jensen, 265–273. Columbus: Ohio State University.

———. 1989. Pilot decision making and judgment. In *Aviation psychology,* ed. R. S. Jensen, 154–175. Brookfield, VT: Gower Technical.

Telfer, R., and Ashman, A. F. 1986. Pilot judgment training: An Australian validation study. Unpublished manuscript. Callaghan, NSW, Australia: University of Newcastle.

Trankle, U., Gelau, C., and Metker, T. 1990. Risk perception and age-specific accidents of young drivers. *Accident Analysis and Prevention* 22:119–125.

Wallston, B. S., Wallston, K. A., Kaplan, G. D., and Maides, S. A. 1976. Development and validation of the health locus of control (HLC) scale. *Journal of Consulting and Clinical Psychology* 44:580–585.

Wichman H., and Ball, J. 1983. Locus of control, self-serving biases, and attitudes towards safety in general aviation pilots. *Aviation Space and Environmental Medicine* 54:507–510.

Wiegmann, D. A., and Shappell, S. A. 1997. Human factors analysis of postaccident data. *International Journal of Aviation Psychology* 7:67–82.

———. 2001. A human error analysis of commercial aviation accidents using the human factors analysis and classification system (HFACS) (technical report DOT/FAA/AM-01/3). Washington, D.C.: Federal Aviation Administration.

———. 2003. *A human error approach to aviation accident analysis: The human factors analysis and classification system.* Burlington, VT: Ashgate.

Wiggins, M., and O'Hare, D. 1993. A skills-based approach to training aeronautical decision-making. In *Aviation instruction and training,* ed. R. A. Telfer. Brookfield, VT: Ashgate.

———. 2003a. Expert and novice pilot perceptions of static in-flight images of weather. *International Journal of Aviation Psychology* 13:173–187.

———. 2003b. Weatherwise: Evaluation of a cue-based training approach for the recognition of deteriorating weather conditions during flight. *Human Factors* 45:337–345.

9 Concluding Remarks

Everyone thinks of changing the world, but no one thinks of changing himself.

Leo Tolstoy

9.1 INTRODUCTION

Notwithstanding Tolstoy's comment, humans are not easy to change. However, after the reader has read this book, we hope that he or she will recognize that, of all the parts of the aviation system, the human is the part most often called upon to change. Fortunately, one of the most characteristic traits of humans is their adaptability—their ability to change their behavior to fit the demands of the situation. There is no more dramatic demonstration of this adaptability than pushing forward on the controls when the aircraft has stalled and the nose is pointing downward—when all one's instincts call for yanking the controls backward.

Nevertheless, humans are not infinitely adaptable. The research on the selection of pilots demonstrates that some individuals are better suited than others. Research on accident involvement also suggests that some individuals are more likely to be in an accident than others—perhaps because they failed to adapt their behavior to the demands of a novel situation. We hope that our readers are now better aware of how the human interacts with the aviation system and has also developed some awareness of the limits of their personal capabilities and adaptability.

By its nature, this book could only scratch the surface of aviation psychology. The topics covered by each of the chapters have been the subjects of multiple books and journal articles. However, the references and suggested readings provided in each of the chapters can lead the interested reader to more in-depth information on the topics. This book, we hope, will have prepared him or her for these readings by providing a basic knowledge of the terminology, concepts, tools, and methods of inquiry of psychology. Building upon that foundation, the reader should now be better able to assess reports that purport to show the impact of some new training intervention or to appreciate the impact of fairly subtle changes in instrumentation design and layout on aircrew performance.

Although we have focused on aviation operations, much of what we have covered is equally applicable to other situations. For example, the task of the pilot has much in common with drivers, operators of nuclear power stations, and surgeons. Poorly designed work stations and controls can contribute to driving accidents and reactor meltdowns as easily as they contribute to aircraft crashes. The principles of designing to meet the capabilities and characteristics of the operators and the tasks they are required to perform remain the same. Only the specifics of the setting are changed.

Along that same line, our focus in this book has been on the pilot; however, we recognize that he or she is only one part of an extensive team. The discussions regarding the pilot also apply to the people outside the flight deck, including mainte-nance personnel, air traffic controllers, dispatchers, and managers of aviation organi-zations. Therefore, what the reader has learned from this book may also prove useful in other settings.

Psychology is concerned with how individuals are alike as well as with how indi-viduals differ. Knowing one's strengths and weaknesses, where one's tendencies can take one, and the limits of one's personal performance envelope can help in avoiding situations where demands exceed capacity to respond.

In the tables that follow, we provide links to a large number of aviation organiza-tions, government regulatory agencies, research centers, and other sources of infor-mation related to aviation psychology or to aviation safety in general. These sites provide a broad range of resources applicable to pilots—from the novice to the most senior airline captain. We encourage the reader to visit these sites to broaden his or her knowledge and acquire new skills. We sincerely hope that readers will use the information from these Web sites and from this book to become more competent and more self-aware, as well as apply what has been learned to being better, safer pilots.

9.2   INTERNET RESOURCES FOR PILOTS

The following tables contain links to the principal aviation regulatory authorities, military services, universities, and other entities related to aviation, aviation safety, and aviation psychology. Some organizations, such as the FAA, AOPA, CASA, and Transport Canada, have many more pages of interest than we have listed here. However, starting from the addresses listed in these tables, readers should be able to locate almost all the relevant material.

Readers are cautioned that although all these links were valid as of February 26, 2009, some of the organizations (particularly the FAA) change the structure of their Web sites without notice and without providing a means to find the relocated materi-als. If that occurs, the reader can try searching for the name of the organization and will probably find the new site.

Be aware that many of the U.S. military sites now have protection systems that make them somewhat incompatible with the more popular Internet browsers (e.g., Internet Explorer). Because of this, a message may be sent to the effect that there are security issues with the site one is trying to reach, and one's browser may issue a prompt to avoid the site. Usually, continuing the operation will result in being taken to the site. It just takes a bit more trust and perseverance.

Civil Aviation Authorities

Organization	Page	Link Address
International Civil Aviation Organization	Home page	http://www.icao.int/
International Civil Aviation Organization	Training on SMS	http://www.icao.int/anb/ safetymanagement/training/ training.html

Federal Aviation Administration	Home page	http://www.faa.gov
Federal Aviation Administration	Aviation manuals and handbooks	http://www.faa.gov/library/manuals/
Federal Aviation Administration	Aviation news	http://www.faa.gov/news/aviation_news/
Federal Aviation Administration	Aviation safety team	http://faasafety.gov/
Federal Aviation Administration	Human factors workbench	http://www.hf.faa.gov/portal/default.aspx
Federal Aviation Administration	Aviation maintenance human factors	http://www.hf.faa.gov/hfmaint/Default.aspx?tabid=275
Transport Canada	Home page	http://www.tc.gc.ca
Transport Canada	Safety management systems	http://www.tc.gc.ca/CivilAviation/SMS/menu.htm
Civil Aviation Safety Authority of Australia	Home page	http://www.casa.gov.au
Civil Aviation Authority of New Zealand	Home page	http://www.caa.govt.nz
Civil Aviation Authority of the United Kingdom	Home page	http://www.caa.co.uk

Accident Investigation Boards

Country	Organization	Link Address
Australia	Transport Safety Bureau	http://www.atsb.gov.au
Canada	Transportation Safety Board	http://www.tsb.gc.ca/
Denmark	Air Accident Investigation Board	http://www.hcl.dk/sw593.asp
France	Bureau Enquetes—Accidents	http://www.bea-fr.org/anglaise/index.htm
Germany	Bundesstelle für Flugunfalluntersuchung	http://www.bfu-web.de/
Ireland	Air Accident Investigation Unit	http://www.aaiu.ie/
Norway	Aircraft Accident Investigation Board	http://www.aibn.no/default.asp?V_ITEM_ID=29
Sweden	Board of Accident Investigation	http://www.havkom.se/index-eng.html
Switzerland	Aircraft Accident Investigation Bureau	http://www.bfu.admin.ch/en/index.htm
The Netherlands	Transport Safety Board	http://www.onderzoeksraad.nl/en/
United Kingdom	Air Accidents Investigation Branch	http://www.aaib.gov.uk/home/index.cfm
United States	National Transportation Safety Board	http://www.ntsb.gov

Other Civilian Government Agencies

U.S. Organization	Page	Link Address
NASA	Aviation Safety Reporting System (ASRS)	http://asrs.arc.nasa.gov/
NASA	Small aircraft transportation systems	http://www.nasa.gov/centers/langley/news/factsheets/SATS.html
NASA	Aircraft icing training	http://aircrafticing.grc.nasa.gov/courses.html
NOAA	Aviation weather center	http://aviationweather.gov/
DOT	Volpe Research Center	http://www.volpe.dot.gov/hf/aviation/index.html
DOT	Transportation Safety Institute	http://www.tsi.dot.gov/

Military Organizations

Organization	Page	Link Address
U.S. Navy	School of Aviation Safety	https://www.netc.navy.mil/nascweb/sas/index.htm
U.S. Army	Combat Readiness Center (safety)	https://safety.army.mil/
U.S. Navy	Air Warfare Center—Training	http://nawctsd.navair.navy.mil/
U.S. Army	Army Research Institute	http://www.hqda.army.mil/ari/
U.S. Department of Defense	Human Factors and Ergonomics Technical Advisory Group	http://hfetag.com/
U.S. Army	Human Research and Engineering Directorate	http://www.arl.army.mil/www/default.cfm?Action=31&Page=31
UK Ministry of Defense	Human Factors Integration Defense Technology Center	http://www.hfidtc.com/HFI_DTC_Events.htm
U.S. Air Force	Human Effectiveness Directorate	http://www.wpafb.af.mil/afrl/he/
U.S. Air Force	Office of Scientific Research	http://www.afosr.af.mil/

University Research Centers

Organization	Page	Link Address
Embry-Riddle Aeronautical University, Florida	Prescott Flight Center Web links	http://flight.pr.erau.edu/links.html
University of North Dakota	Resources	http://www.avit.und.edu/f40_Resources/f2_Podcasts/index.php
Cranfield University, United Kingdom	Aerospace research, including psychology	http://www.cranfield.ac.uk/aerospace/index.jsp

Ohio State University	Aviation Department	http://aviation.eng.ohio-state. edu/
Arizona State University	Cognitive Engineering Research Institute	http://www.cerici.com/
University of Texas	Human Factors Research Project	http://homepage.psy.utexas.edu/ homepage/group/ HelmreichLAB/
University of Tromsø, Norway	University Research Center	http://www2.uit.no/www/ inenglish
George Mason University, Virginia	Center for Air Transportation Systems Research	http://catsr.ite.gmu.edu/
University of Otago, New Zealand	Cognitive Ergonomics and Human Decision Making Laboratory	http://psy.otago.ac.nz/cogerg/
Trinity College, Dublin	Aerospace Psychology Research Group	http://www.psychology.tcd.ie/ aprg/home.html
University of Illinois at Urbana-Champaign	Institute of Aviation	http://www.aviation.uiuc.edu/ aviweb/
National Aerospace Laboratory NLR, The Netherlands	NLR home page	http://www.nlr.nl/
Maastricht University, The Netherlands	University Research Center	http://www.unimaas.nl/http:// www.unimaas.nl
Monash University, Australia	Accident Research Center	http://www.monash.edu.au/ muarc/
University of Graz	International summer school on aviation psychology	http://www.uni-graz.at/isap9/

Organizations

Organization	Page	Link Address
AOPA/ASF	Home page	http://www.aopa.org/asf/
AOPA	Pilot training	http://www.aopa.org/asf/ online_courses/#new
AOPA Air Safety Foundation	Weather training	http://www.aopa.org/asf/ publications/inst_reports2. cfm?article=5180
Flight Safety Foundation	Aviation Safety Network	http://aviation-safety.net/index. php
Flight Safety Foundation	Links	http://www.flightsafety.org/ related/default.cfm
Experimental Aircraft Association	Home page	http://eaa.org/
National Association of Flight Instructors	Home page	http://www.nafinet.org/
Aviation Safety Connection	Home page	http://aviation.org/
Austrian Aviation Psychology Association	Home page	http://www.aviation-psychology. at/index.php

European Association for Aviation Psychology	Home page	http://www.eaap.net/
UK Royal Aeronautical Society	Home page	http://www.raes-hfg.com/
APA Division 19 Military Psychology	Home page	http://www.apa.org/divisions/div19/
APA Division 21 Experimental Psychology	Home page	http://www.apa.org/divisions/div21/
Australian Aviation Psychology Association	Home page	http://www.aavpa.org/home.htm
International Test Commission	Standards for psychological tests	http://www.intestcom.org/
Association for Aviation Psychology	Home page	http://www.avpsych.org/

Other

Organization	Page	Link Address
American Flyers	Provides access to FAA videos	http://www.americanflyers.net/Resources/faa_videos.asp
International Symposium in Aviation Psychology	Meeting held every other year devoted to aviation psychology	http://www.wright.edu/isap/
U.S. Army Research Laboratory	Helmet-mounted displays in helicopters	http://www.usaarl.army.mil/hmd/cp_0002_contents.htm
International Military Testing Association	Home page	http://www.internationalmta.org/
American Psychological Association	Information on tests and test development standards	http://www.apa.org/science/testing.html
Norwegian Institute of Aviation Medicine	Aeromedical research	http://flymed.no
Office of Aerospace Medicine of the FAA	Technical reports produced by the FAA on aviation psychology	http://www.faa.gov/library/reports/medical/oamtechreports/
U.S. Civil Air Patrol	Safety	http://level2.cap.gov/index.cfm?nodeID=5182
Neil Krey	CRM developers' page	http://s92270093.onlinehome.us/CRM-Devel/resources/crmtopic.htm
Aviation Weather.Com	Aviation weather maps	http://maps.avnwx.com/
National Weather Association	Weather courses	http://www.nwas.org/committees/avnwxcourse/index.htm
International Journal of Applied Aviation Studies	Scientific articles on aviation psychology topics	http://www.faa.gov/about/office_org/headquarters_offices/arc/programs/academy/journal/

University Corporation for Atmospheric Research	Meteorology education and training	http://www.meted.ucar.edu/
Quantico Flying Club	Weather training	https://www.metocwx.quantico.usmc.mil/weather_for_aviators/pilot_trng.htm
Aviation Human Factors[a]	Home page	http://www.avhf.com
SmartCockpit	Large aircraft safety issues	http://www.smartcockpit.com/
Airbus	Safety library	http://www.airbus.com/en/corporate/ethics/safety_lib/
U.S. Department of Defense	Human Systems Information Integration Analysis Center	http://iac.dtic.mil/

[a] In the spirit of full disclosure, it should be noted that this site is maintained by one of the authors.

Linkage Sites

Organization	Page	Link Address
Flight Safety Foundation	Links to aviation sites	http://www.flightsafety.org/related/default.cfm
Landings Web Page	Aviation safety links	http://www.landings.com/_landings/pages/safety.html
NodeWorks	Links to other sites	http://dir.nodeworks.com/Science/Technology/Aerospace/Aeronautics/Safety_of_Aviation/
Human Performance Center—Spider	Links to other sites	http://spider.adlnet.gov/

Readers will have noted that all of the preceding sites are English language sites. Of course, the accident investigation boards will have parallel native-language sites, as do sites such as the NLR. Undoubtedly, many other non-English sites are also available; however, for various reasons a majority of the research and publications in aviation are in English—hence, the preponderance of Web sites in the English language.

Index

A

Accident(s), 9
  aircrew errors in, 184
  causes, 178–181, 182
    swiss-cheese model of, 184, 185
  classification, 181–187
    mechanical and maintenance-related, 181, 182
    pilot-related, 181, 182
  close calls and, 188–191
  decision-making component, 191–193
  fear of flying and, 141–142
  global information network, 10
  human error approach to analysis of, 175
  incidence, 177–178
  investigation boards, 213
  mechanical and maintenance, 182
  model of causation, 11, 111
  multiple layer concept, 110
  Nall report, 70, 182
  national culture and, 157
  organizational issues associated with, 153–154
  pilots and, 12, 191
  potential, 189–190
  psychological test scores and predictability of, 5, 16, 54
  research on, 187–191
  risk, 137–138
  sleep and, 6
ACT-R, 11
ADM. *See* Aeronautical decision making (ADM)
Advanced beginner, 10
Advanced Qualification Program, 101
Aeronautical decision making (ADM), 193
  effectiveness of training, 193–194
  models, 12–17, 18
    DECIDE, 12
    expertise, 14
    Hunter's, 13–15
    of pilot judgment, 13
    safety initiatives, 208
Age of effect, 133
Air France Airbus A-320, 59
Air rage, 146–148, 149
  causes, 147–148
  defined, 146
  incidence, 146–147
  investigation, 146
  prevention, 148
Air traffic controller(s)
  job analysis, 75
  selection, 95
    assessment center for, 77
    biographical data and, 77
    historical overview of, 85–86
    interview for, 76
    meta-analysis results for, 87
    predictors in, 76–78
      evaluation of, 79–83
    psychological tests for, 77
    school grades and, 77
    validity of methods for, 78, 79–83
    work experience and, 77
    work sample test for, 77
Air travel, pains and pleasures, 144–145, 151
Aircraft configurations, control confusion and, 52, 53
Altimeter reading, 54
AS-LOC. *See* Aviation safety locus of control (AS-LOC)
ASRS. *See* Aviation Safety Reporting System (ASRS)
Associative phase of skill acquisition, 10
ATCO. *See* Air traffic controller(s)
Autonomous phase of skill acquisition, 10
Aviation, civil authorities, 212–213
Aviation psychology
  associations, 215–216
  defined, 1–3
  research models, 8 (*See also* Model(s))
  system design and, 62–64
    current issues of, 65–69
Aviation safety. *See* Accident(s); Safety
Aviation safety locus of control (AS-LOC), 197
Aviation Safety Reporting System (ASRS), 68
Aviation weather encounters, 201–202

B

Beginner, advanced, 10
Burnout, 31, 130–132, 170, 173. *See also* Stress
  banishing, 150
  cynicism and, 131
  defined, 130
  depersonalization and, 130
  emotional exhaustion and, 130
  exhaustion and, 131
  inventory manual, 150